W9-CEE-286

Praise for *Before the Streetlights Come On*

"Climate change affects all of us, but it doesn't affect us equally. All too often, those most affected are already overwhelmed by the cascading impacts of inequity and injustice. Drawing from her vivid life experiences and wealth of knowledge, Heather McTeer Toney sounds a clarion call for immediate climate action in and for marginalized communities. Why? Because if we don't fix climate change, we can't fix anything else."

—**Dr. Katharine Hayhoe**, climate scientist and author of *Saving Us: A Climate Scientist's Case for Hope and Healing in a Divided World*

"With crystal clarity, Heather McTeer Toney documents the shameful history that has left Black communities particularly vulnerable to the ravages of climate change—and explores what we can do about it."

—**David Axelrod**, director of the University of Chicago Institute of Politics, and CNN senior political commentator

"Now is a time for deeper and more diverse public thought about climate and environment. Heather McTeer Toney is taking up the challenge."

—**Tamara Toles O'Laughlin**, national climate strategist and founder of Climate Critical Earth

"*Before the Streetlights Come On* breaks climate out of its wonky shell and brings it into the living room, backyard BBQ, and fellowship hall. Using the language of everyday people, Heather McTeer Toney blends common sense, humor, and storytelling. This book will welcome you into the topic or help you see it with fresh eyes."

—**Dr. Katharine K. Wilkinson**, coeditor of *All We Can Save: Truth, Courage, and Solutions for the Climate Crisis*

"McTeer Toney brilliantly describes the real impact and politics of climate change at the community level. If you want to understand the way forward—the way we solve this challenge through the democratic process—you need to read *Before the Streetlights Come On.*"

—**Fred Krupp**, president of the Environmental Defense Fund

“Heather McTeer Toney gives us a practical, accessible, essential book that directly responds to Mother Earth’s call: ‘Say my name; you gonna respect me,’ as she dishes out some ‘Do Right’ to make sure we all know she’s pissed about how we have disrespected the gift and how she expects us to do better moving forward.”

—**Mustafa Santiago Ali**, EVP, National Wildlife Association

BEFORE THE STREETLIGHTS COME ON

BEFORE THE STREETLIGHTS COME ON

Black America's Urgent Call for Climate Solutions

Heather McTeer Toney

Broadleaf Books
Minneapolis

BEFORE THE STREETLIGHTS COME ON
Black America's Urgent Call for Climate Solutions

Cover design: Leah Jacobs Gordon

Print ISBN: 978-1-5064-7862-3
eBook ISBN: 978-1-5064-7863-0

For Deriah and Devin
May you forever make our ancestors proud

CONTENTS

CONTENTS

FOREWORD

Over the decades we have said a lot about climate, the environment and the harrowing impact on Black people across the country. In fact, between the two of us we have written hundreds of thousands of pages, researched and taught well over sixty years of material and spoken countless words of warning to shed light on the dangers placed upon environmental justice communities when these issues were not addressed. Yet today we are still fighting like never before to make sure those on the frontlines of the climate crisis are provided the voice and resources not only to combat the threat of climate change, but also to recover from past injustices while preparing a more equitable future for generations to come.

That's why this book is so important. Heather McTeer Toney brings to life the intersection of climate change and the social justice issues faced by Black Americans daily through her stories and lived experience. It is a refreshing account of the facts about climate change and environmental injustice mixed with our ongoing social and civil rights struggle. Climate change affects every facet of Black life. Whether it's the tragic killing of Black men at the hands of the police, economic disparities in Black communities or healthcare, it can all be linked to climate change and the disproportionate impacts to poorer, often Black, communities. Filled with history, humor and hope, Heather takes us on a journey of real-life encounters of everyday people with climate change and environmental injustice. Through her eyes we see ourselves in climate change—our families and friends—while also envisioning a glimpse of what future climate solutions could be by valuing our input.

We've known Heather as a student, a local elected official, a federally appointed official under the Obama administration, a climate activist, a mother, a mentee and a friend. So it doesn't surprise us to see this book. Like us, she is from the American South. It is a place steeped in historical barriers for Black Americans but also the birthplace of some of the fiercest social justice activists and change makers in the world. Heather's work is continuing the tradition of education and communication through storytelling, and we are proud that the next generation has not forgotten the path of resiliency forged by our ancestors.

We wrote *The Wrong Complexion for Protection: How the Government Response to Disaster Endangers African American Communities* from our real-time, real-life experience as Black people who also happened to be scholars that survived some of the most devastating disasters in the United States, including environmental and public health emergencies, toxic contamination, oil spills, flooding and hurricanes. We also examined post-disaster resource allocations through a racial equity lens and surmised the truth that we've known and experienced our entire lives—place matters. In order to address the prevention, precaution and avoidance of harm, we must work to eliminate unfair, unjust and racist policies and practices. This is the urgent solution we need in order to effectuate equitable climate policy. Heather's book will help people of all backgrounds see clearly the effects of unjust and unbalanced actions on Black America as well as understand why it is critical to our collective survival. We hope you read, share and enjoy it as much as we do!

—*Drs. Robert D. Bullard and Beverly Wright, authors of* The Wrong Complexion for Protection: How the Government Response to Disaster Endangers African American Communities

CHAPTER 1

HURRY UP AND GET HOME BEFORE THE STREETLIGHTS COME ON!

> *For we have built into all of us, old blueprints of expectation and response, old structures of oppression and these must be altered at the same time that we alter the living condition which are the result of those structures. For the master's tool will never dismantle the master's house.*
>
> —Audre Lorde

How in the world are Black folks supposed to talk about climate change when we have other pressing issues to deal with? How and better yet, why?

I've asked myself these questions a thousand times. I hear them when I talk to Black people about climate and environmental topics. Why talk climate in an era of George Floyd, Ahmaud Arbery and Breonna Taylor; with BBQ Beckys and Karens monitoring Black behavior; amid pandemics like COVID-19; while school shootings and blatant voter suppression continue unabated? How in the hell do we have time to think about climate change? What about it should divert our attention from these real and pressing concerns?

When I ask Black people to describe climate change—to define climate change in relation to Black life—my closest friends and family look at me like, "Heather, you are acting like the stereotypical Black girl named Heather. That's white people stuff. If the planet burns up, at least my student loans will be gone. Don't act like you didn't have to be in the house before the streetlights came on like the rest of us. We got other shit to worry about. Prioritize, sis. You need a visit back home to Greenville, Mississippi."

I chuckle to myself because it *is* about home. Climate change is so much about our homes, families and loved ones that we need to hurry up and fix it so that we can all get home safely. I was born and raised in Greenville, Mississippi, the heart and soul of the Mississippi Delta. While my parents were Northerners swept south to the call of civil rights advocacy in the early 70s, my brother and I are natives of Delta soil. We were wrapped in the familial love of people who counted struggle together just as important, if not more, as blood relation. When I think of home, I think of my uncle Willie Bailey, my dad's law partner, who would come to the house and sneak my grandmother Wild Turkey whiskey when my dad was away. I think of my sisters and brother from another mother, Trina, Valerie and Eric Gannison. Their mother, Mrs. Ethel Gannison, was one of my mother's cherished friends. I wouldn't stay with a babysitter without Valerie. Trina would plait my hair with beads and foil on the ends. Eric and I would race up and down Delta Street in front of their house until the streetlights came on.

When I think of climate change in relation to Black life, I relate it to streetlights. Like a lot of Black children, regardless of generation, I had to be in the house before the streetlights came on. City or suburban, Black folks know what "before the streetlights come on" means. In Black neighborhoods, the determiner of time was not a watch or cell phone. A friend's mother never yelled out the back door, "Be home by five o'clock sharp!" A specific hour and minute were too easy to ignore.

But the streetlight? The streetlight was a common denominator that gave both time and distance boundaries. The streetlights coming on meant it was about to get dark and light was required to safely navigate the neighborhood. The streetlights meant that all play action needed to be wrapped up. Foot races had to end. Bicycles had to be put up. You'd best be well on your way home BEFORE the lights came on. In Greenville, Mississippi, streetlights were on big wooden poles and glimmered as they went from dim to fully lit. That took about thirty seconds. That was a heads up: dash home. When I played with my cousins in Columbia, Maryland, I remember one big streetlight on a tall metal pole at the end of the street. That sucker just popped on and *voilà!* We were late. No matter the location, the language of the streetlight timer was the same. Friends didn't let friends stay out past when the streetlights came on. When one had to be in, all had to be in. I am my sister's, brother's, cousin's, next-door neighbor's and church member's keeper.

Just because there wasn't a city-operated light atop a tall pole didn't mean there wasn't a streetlight. Streetlights came in all shapes and sizes. Streetlights indicated boundaries. Different communities had different kinds of streetlights. When the barbershop on the corner flipped on the light in the window, that meant it was about to get dark, head home. Every streetlight didn't always work. If you were in an area where ALL the streetlights were ALWAYS out, you didn't have any business there in the first place!

Streetlights had the power to make you reconsider your whole entire life library of choices. Times I considered whether or not I could stay out past when the streetlights came on, I heard in my head words uttered from the lips of my own mother, the pastor's wife at our church or one of my play aunties: *"I know that you know better." "Your actions clearly indicate you forgot who you are." "Get yourself together before you disparage the family name, child." "Your mama didn't raise no fool." "Use the common sense God gave you."* These words pulled me back into reality from whatever foolishness I was about to pursue as the streetlights were coming on. The voices of elders spurred an inner call to action, an immediate reminder that deep inside I had the wherewithal to make the right decision.

We could control what we did in our world before those lights turned on. But when they did, "Hurry up, we ain't got all day!" was the call to action. Right now, that same call to action is carried in the waves of massive hurricanes, on the winds of devastating firestorms and in the uncharacteristic heat of winter. Climate conditions have escalated past the point of danger. We need the world, including the Black community, to respond accordingly. Black Americans are steeped in a history of collaboration with nature. Our cultural inheritance of overcoming any odds against us and our sheer will to survive makes us well-suited leaders in the collective global response.

To me, as a climate solutions advocate, the phrase "the streetlights are coming on" perfectly encapsulates the way Black people encounter and address climate and environmental injustice in our country. I make no assumption that I represent the thoughts, ideas and patterns of all Black Americans. We are not one monolithic group. The diversity of Black American people is part of what makes us special. Still, there is a common thread among us, a link that connects us to the environmentally protective wisdom of our ancestors. It is almost second nature. Even when we're oblivious to this aspect of our cultural inheritance, we often use this environmentally protective wisdom to solve problems we encounter. No one has to tell us—we just know.

HISTORY BEHIND THE "WHY" WE MUST ADDRESS CLIMATE CHANGE

In the Black community, we have other things to worry about besides climate change. The idea gets lost in the cloud of issues we muddle through daily. When listed next to job security, food insecurity, gun violence and the blatant racism faced daily by African Americans, climate change ranks low among problems competing for our attention.

But as African Americans, we are disproportionately impacted by the effects of climate change. Black people make up 13 percent of the US population, but breathe 40 percent more dirty air than our white counterparts. We live in areas four times as likely to be impacted by hurricanes, tornadoes and floods, and we are twice as likely to be hospitalized or die from climate-related health disparities.

Surviving traumatic change is part of our lived history, not new to our experience. Imagine the upheaval and confusion enslaved Africans experienced as they were stolen from their homes and forced to relocate and adapt in new environments across the New World. Most were taken to locations in the global south like Brazil and the Caribbean while others landed in the continental United States. We are descended from people traumatically removed from one ecosystem—the linked biological community of plants, animals, people, germs, energy—and transported across the Atlantic Ocean to an entirely different ecosystem. Under the most horrendous and inhumane conditions imaginable, the enslaved were forced to teach and care for people from another ecosystem—European—and to figure out how to make plants grow and animals thrive. My ancestors brought with them skills they quickly learned to apply to the land they were forced to tend. If there is any group of people that can innovate climate solutions and adapt, it's Black Americans. If our ancestors figured out how to survive the hot underbelly of a ship and adapt to a wholly different environment, we can also figure out how to handle the climate crisis. Our history underscores the value of Black people's role in the climate movement. We know how to adapt to change.

I love the example of recycling and reuse. Recycling isn't a "new" method to reduce plastics and waste in Black households. Enslaved Africans creatively recycled and reused every item they encountered. Our great-grandmothers repurposed leftover materials to create beautiful quilts, patterned with the stories of struggle and survival. They wrapped us in the warmth of their love and legacy. Scraps of meat and vegetables were turned into succulent dishes prepared with care and prayers for nourishment. A plastic bag from the grocery store was also a trash bag, hair conditioner cap, lunch box, Halloween bucket, stuffing for mailing breakable items and what you wrapped the lotion bottle in when traveling so it wouldn't spill on clothes. For us, recycling and reuse isn't just to protect the planet. It is a way of life, a nod to our memories, a way to protect what we had and keep what we have. Today, these recycle and reuse lessons remain in our culture regardless of how much money we have. While it wasn't right, we managed to survive historical climate and environmental injustices while addressing the multitudes of social justice issues plaguing minority and often marginalized people. This is one example of the ways we have naturally responded to climate crisis. It's streetlight security at its finest.

Climate and environmental issues have always been intertwined with our struggles for justice. But now the streetlights are on and we must see clearly. Trust me, making sure there are equitable climate solutions that speak to the experience of all people is tantamount to running home before those streetlights glow.

Black academics, community leaders and scientists who work in environmental and climate issues don't get a break from other injustices that impact Black America. Working in climate doesn't make us immune to the varied injustices that hurt our sons and daughters. After talking about climate change, I go home to concerns of my husband being pulled over by the police or my sweet five-year-old son being categorized as too aggressive when he's playing. Before giving a speech on environmental justice, I worry that my daughter has to deal with bullying because she's in a majority-white school and people either pick on her stunning African features or question her Blackness because of her brilliance. The multitude of social justice issues that weigh on Black people never gives way to one or the other, nor to environmental and climate injustice.

Despite the myriad of persistent struggles faced by Black American climate advocates, we still press the focus on the environment and climate change. We do this because the science is clear—the climate has changed and continues to do so now. Devastation is taking place as we speak. Time waits for no one. Climate degradation and environmental injustice are deadly factors in Black communities, not unlike killer cops and uncontrolled access to firearms.

The difference between mainstream majority-white environmental movements and minority-led Black, brown and Indigenous environmental movements is that the latter does not have the luxury of silo. Our issues coexist. Climate change collides with other historic and systemic racially based issues to create a long-overdue desire for one thing: equity.

We've all heard the age-old colloquialism that we're all in the same boat, so we must row together and in the same direction if we're going to survive. But that's not necessarily true. When it comes to climate and environmental issues, humanity is in the same storm, but we are not in the same boat. Some of us are sailing along in yachts or manning aircraft carriers while others are bobbing along in rafts and rowboats. The impact of the storm hits us differently. The one thing we have in common is that we all want to survive.

Our foremothers and forefathers made it the business of the village to ensure that the next generation understood the importance of doing our best to keep everyone safe. Our history serves as a clarion call to environmental consciousness. African Americans have always been molded or influenced by our environment. In our recent history, this one simple phrase "before the streetlights come on" is our prompt to participate in the restoration and healing of the air, land and water in our communities and we need to hurry up.

SURVIVAL OF THE WORST ENVIRONMENTAL CONDITIONS IMAGINABLE

In 2012 I traveled to Dakar, Senegal, with a group of Black mayors, faith leaders and journalists from the United States. As guests of the president, we were granted private access to Gorée Island, Dakar's oldest and largest slave house. I was ill-prepared for the wave of emotions, images and physical impressions that flooded my body as we approached the castle and docked to disembark. An unrecognizable fear gripped me as I stepped off the boat and the guides led us into the courtyard of the fortress. I wondered to myself, is this what my ancestors felt? As a free Black woman, citizen of the United States of America and holder of a passport, I looked over my shoulder for fear that I may somehow be cast back to the period of enslavement and forced into the experience. As I walked from hold to hold, I stepped gently into the space that once held women, children and men. The noticeable lack of airflow was immediate. I imagined being taken from grasslands where the wind freely flowed across my face and stuffed into a small hole with little oxygen to spare. Seeing and touching the iron chains that bound human captives as they were prepared for enslavement sent electric shockwaves through my skin. The rust color of the iron bore such resemblance to the rust color of the floor that, unconsciously, my mind forced my other senses to engage—I couldn't help but smell and taste captivity. Is this what it felt like to, inexplicably and without control, be thrown into a different environmental space?

At the back of the fortress is a rectangular stone opening with a wooden door. Beyond it lies the ocean. This is known as the "Door of No Return." As I approached it, tears streamed quietly down my cheeks. So many had traversed through this door into a world yet unknown to them and unyielding in its cruelty. Was I feeling the deep despair of being thrust into a new atmosphere without consent? In the sound of the waves crashing on the rocks below, I could hear a still, small voice. The violent, constant crashes brought a repetitive melody that somehow brought me peace. I could hear the sing-song voice proudly say, "Ah! But they were mistaken! We have returned! We were not forgotten! We did survive and our legacy lives on through you." In that moment and spirit of hope, I wiped my face, recognizing the breadth and scope of sacrifice. If they could face the unknown future of a different environment and atmosphere, then I have a duty to ensure that their fight to survive is not in vain.

If there's any group that can survive and adapt to a changing climate and environment, it's Black Americans. After being boarded onto a ship with up to four hundred other humans, captives were made to lie prone, with roughly six to eight feet of space from head to toe. On average, the Middle Passage voyage lasted eighty days. That's equivalent to over two months without access to clean air, water and land. The inability to breathe fresh air during this passage was arguably the most deadly. The human lungs are designed to push out toxins and stale air while replenishing the body with fresh oxygen. In doing so, the basic functions of our organs—pumping blood, digesting food, fighting bacteria—are able to do so efficiently while giving our brain the energy to think clearly and our muscles the ability to move freely. Access to clean, fresh air greatly reduces the transmission of virus and disease. Closed spaces with no airflow are breeding grounds for sickness. Since there was little to no ventilation in the slave hold, the circulated air that held feces, blood and urine, combined with saltwater and human flesh, created an environment for the rapid spread of virus and disease.

Inflammation of the lungs made up over half of the causes of death at sea during the transatlantic Middle Passage period. Death was daily, either through sickness, suicide or the simple unwillingness to go on. Many that survived would carry both the physical scars of the voyage as well as the epigenetic scars that were passed down from generation to generation. It's no surprise that African Americans are among the highest percentage of minorities to suffer disproportionately from respiratory illness. The fact that my great-great-great-great-grandfather was taken from an environment of open fresh air and forced to spend three months in an unventilated, virus-filled slave hold is directly related to the fact that two of my three children may suffer from asthma and my grandma might end up with COPD. In a study entitled "Chronic Obstructive Pulmonary Disease in America's Black Population," researchers found that while studies have shown the lung function of Black adults and children is lower than white adults and children, there is an additional difference between nineteenth-century immigrants from Black countries, such as Jamaica, Nigeria and Kenya, and those considered US-born descendants of enslaved Africans. The lung function of those born in the United States was lower than our ethnic cousins. Combined with contributing socio-economic factors, such as access to healthcare, insurance, substandard housing, poverty and education, African Americans have been gasping for fresh air since 1619.

Mere survival of the transatlantic slave trade should be evidence enough; however, the resilience of enslaved Africans through warmer years that made the journey harder makes it more evident. The hurricanes that bombard and cause devastation across the global Southeast are formed off the west coast of Africa. The majority of hurricanes formed in West Africa travel the same path as slave ships during the period of the Middle Passage. While there are many African and Black American folktales that imagine hurricanes as angry ancestors carrying their revenge across the Atlantic while gathering the spirits of those cast overboard during the journey, the reality is that hurricanes and slave ships are driven by the same element—trade winds. These global winds move east to west off the coast of Africa and are part of our global weather system.

Warm waters weren't the only thing to make the trip difficult. Extreme heat and drought were also part of the equation. Environmental factors and climate fluctuations dating back to 1730 show correlation among the profit, temperature and mortality of the slave trade. In warmer years, crops yielded less food and the cost of feeding captives and stocking ships increased. Hot weather also meant that viruses spread more easily, contributing to higher rates of disease and mortality. These factors resulted in not only a rough ride, but a limited number and higher transportation cost for enslaved Africans doomed to make the journey. To survive any of this is akin to a miracle.

Today, there are parallels between the enslaved experience of our ancestors and the current social impacts of environmental injustices and climate change in communities. It's striking that housing projects in urban Black America were fashioned to stack families one on top of the other—closed spaces with little to no ventilation, windows barred shut by iron with no access to fresh air, stifling elevators and stairwells etched with the stench of urine. The lived experience of many African Americans was to dwell in an uncanny likeness to a modern-day slave ship. More unresolved inequity is revealed when we look at where Black communities are geographically located throughout the United States. Due to where enslaved Africans were brought to work, over half our nation's Black population live in the warmest part of the country: the South. As temperatures rise and vector-borne diseases spread, our communities are disproportionately impacted. Black communities are hit first and hardest from hurricanes and floods, yet are given inadequate resources to repair before the next storm. The community of Port Arthur, Texas, a majority-Black town located along the Texas petrochemical corridor, is known as Hurricane City. In 2020, during the height of the coronavirus pandemic, Port Arthur was hit by Hurricane Harvey. Less than a month later, Hurricane Laura arrived. Families barely made it back into their homes to recover before the next storm barreled towards them. Even facing repeated climate-related incidents, they proudly survive.

Not only did our ancestors survive, but they taught. While their European oppressors—most of whom were also from a different continent and ecosystem—struggled to adjust, the wisdom of enslaved Africans became instruction. By way of agricultural knowledge and farming practices gained in their native African countries, the enslaved taught those enslaving them how to gain greater crop yields from rice, sugar and tobacco. These survivors were relied upon and invaluable assets both in valued sense of knowledge and the immoral sense of commodity.

Evidence of African American connections to the environment became stronger with the expansion of land-grant colleges for Black students. The 1890 land-grant institutions program established state-affiliated Black colleges and universities that focused on research in the areas of agriculture, forestry and food. Most historically Black colleges and universities were founded under this program and continue to develop innovations around renewable energy, biotechnology and food safety. The support of Black colleges also helped solidify the beginning of the environmental justice movement in America. In 1982, students protested alongside residents of Warren County, North Carolina, against the permitting of a landfill in a Black community. But long before that historic protest associated as the beginning of the environmental justice movement, the Black, youth-led Student Nonviolent Coordinating Committee (SNCC) was founded in North Carolina under the leadership of the fearless and relentless Ella Baker. The granddaughter of an enslaved woman who was beaten for not marrying the man her owner selected, Ms. Baker embodied resilience and the determination to face injustice head on. After helping Dr. Martin Luther King Jr. organize the Southern Christian Leadership Conference and coordinating the Woolworth lunch counter sit-ins, Ms. Baker met with students at Shaw University to start SNCC and was a force of advocacy for social justice issues including environmental racism.

All of the facts, history and stories beg the question: Why aren't African Americans sought after, let alone better engaged in conversations about climate solutions?

Did it take a pandemic to reveal that Black people become sicker and die faster because we are disproportionately exposed to air pollution? The housekeepers and janitors in South Louisiana could have told you that back in the early 1980s. Why did it take watching the police kill George Floyd as he was screaming, "I can't breathe!" for us to realize that demanding justice in every breath is a demand for the eradication of both the air pollution we breathe and the violence we face? Any mother could have explained that eloquently. Our roots in America run deep through the physical soil by way of enslavement and sharecropper work of the land. Through time, we expanded these proficiencies through higher education, innovations and environmental social justice movements. Our legacy of survival should—at minimum—grant a communal trust to Black environmental experts.

Dr. Robert Bullard is often referred to as the "Father of Environmental Justice." I am blessed to know him as his student, a colleague and a mentor. A nonsensical, sturdy man with a hell of a sense of humor, Dr. Bullard has no problem telling anyone exactly what's on his mind and detests when unfamiliar people refer to him as "Bob." He's authored over nineteen books, received too many awards to count and has spoken at countless conferences, congressional hearings and seminars. While attending a Society of Environmental Journalists conference, we stepped away to grab lunch and he shared a glimpse from his years of experience. Dr. Bullard said, "Heather, you know years ago I would show up to a conference I'd been invited to, and the conference leaders would take one look at me and claim my invitation was somehow a mistake. They couldn't envision a Black man, PhD as an environmental expert. It's changed, but now folks are worried about Critical Race Theory when they should be worried about critical breathing." I thought to myself, "Ain't that the truth!" If not for the lived experiences of experts like Dr. Bullard, we miss the real issues and lose the people-impact of our ideas. These experts should be one of the first stops when looking for solutions.

Black, brown, Indigenous, marginalized and impoverished people around the world continue to demand that their voices are included in culturally competent and responsive solutions to climate change. 2020 was one of the hottest years on record, creating a myriad of devastation that compounded the physical, mental and economic stress felt by the Black community. Coastal cities and towns were ravaged by a record-breaking hurricane season while farmers experienced crop devastation. As I began writing this book, the coronavirus disease has killed over 613,000 Americans with Black and brown communities suffering the highest percentage of infection and death. We're up to 813,000 and rising.

The hardest-hit communities continue to display a tenacity to thrive. When I see Black communities rebuilding after a climate-related or extreme weather event, I envision a collective response resounding in that common phrase: "We're not new to this. We're true to this," an epic clap back exemplifying the resiliency of a people.

Climate change is a voter issue, an economy driver and a defining dynamic for the foreseeable future—and it deeply affects the Black community. Without a global awareness and understanding of why we should care, what should be done and our unique experience to create solutions for everyone, the world risks excluding the participation and possible solutions from an influential part of our society. Awareness of the role of climate issues on the Black American experience today is like the streetlights coming on. Whatever else we are doing, we need to bring this awareness home because no matter what we call it—climate change, climate crisis, climate safety, climate justice—it all affects Black communities. We are losing daylight by not responding right now.

Throughout the pages of this book, it's my sincere hope that you discover the deep connections between the culture and experience of African American people, how the environmental and climate issues have impacted every facet of Black life, and how we are well-suited to be drivers of the solutions. I cannot and would not attempt to speak for all Black people. The references to "Black folks, Black people and the Black community" are merely my lived experiences of growing up, working and serving in a number of capacities that have allowed me to see the multitude of conversations within many aspects of African American life. I am a self-proclaimed recovering politician. I was elected as the first African American, first woman and youngest mayor of Greenville, Mississippi. I have led amazing organizations from the National Conference of Black Mayors to being part of the first cohort of Young Elected Officials, a program that counts United States cabinet members and senators among our alumni. I was appointed as the Regional Administrator for the Southeast Region of the Environmental Protection Agency (EPA) under President Obama, where

I oversaw the federal environmental regulations of eight states and six federally recognized tribes.

Now in my role as Vice President of Community Engagement at the Environmental Defense Fund, I guide co-created community driven solutions to climate change. Throughout my career as a local elected official, a federally appointed administrator and environmental activist, I have not seen as much hunger for this information as I do now. Extreme weather, coronavirus disease, and murder hornets are just a few of the "end times" events that raise the eyebrows of my friends, family and neighbors in the Black community.

This book addresses climate-related impacts in the Black community through the lens of social justice issues. Everything from gun violence to education to our neighborhood air. It's also a book to empower us to talk about solutions. But this is not a book just for Black people. The climate crisis is a global existential emergency—so goes one, so goes us all. It behooves everyone to embrace the intersections of climate to each other as well as the major justice issues plaguing marginalized people if we are to develop solutions that empower marginalized people to act. It's time to look at what has hindered cohesive engagement and partnerships. It's time to bring frank discussion and resources to climate action in our conversations and advocacy starting at home.

We need more Black voices and leadership on issues of climate change and environmental justice. The streetlights are turning on, indicating climate warnings that cannot be avoided. Traditional green groups must share power and resources as we rebuild trust and work on community-based climate solutions together. Systemic racism has benefited most industries, but no one wins when diverse voices are excluded from protecting the planet.

In this book, you won't find packaged answers. What you'll find are narratives and stories about climate and environmental action that have proven effective to the survival of Black communities. We can aid humanity by making the climate crisis visible through our eyes and normalizing climate conversations as part of achieving equity. As we tackle the challenges that life presents to us all, I am reminded of the brilliance, wisdom and ingenuity it took for my ancestors to assure my very existence. We must embrace our culture as a way to speak to the urgency of the crisis. And guess what? If we hurry up, we can get there before the streetlights come on.

CHAPTER 2

THE BASICS OF CLIMATE CHANGE, IN PLAIN LANGUAGE

Until dear Mother Nature says her work is through
Until the day that you are me and I am you

—Stevie Wonder, "As"

"Stop running in and out of this house. Y'all lettin' all the good air out!"

The number of times I've heard the phrase "Y'all lettin' all the good air out!" while I was growing up in the Mississippi Delta would easily qualify for platinum status if it were a song.

During the summer months of the early 1980s, our next-door neighbor Katie—often still wearing her pajamas—started the day early in the morning by sitting on the concrete back steps to our house, waiting for my brother Marcus and I to come out and play. We were the only kids on our block of Washington Avenue, a busy main thoroughfare in the historically white part of Greenville, Mississippi. It did not matter that we were Black and she and her younger brother, Daniel, were white. We were the only four kids who lived on our street, and nothing was more important than play. From sunup to sundown, we played inside our houses and outside, rain or shine. It didn't matter what other activity was sacrificed or whose house we were running in and out of, play was our priority. We'd play outside during the hottest time of day. When exhausted, we'd take a break, cool off, regroup and get back to play.

Why in the world did it matter if we left the back door cracked? Who was it going to hurt? We had no concept of the cost of electricity, the wear and tear on a constantly running air conditioner, or the fact that flies and mosquitos got inside and hid under our beds, waiting to bite us at night. And for goodness sake, could somebody please explain how one lets "good" air out?

The frustration of our parents and caretakers grew steadily as we ignored repeated warnings. We felt consequences would never come and their yelling was just rhetoric to draw us away from our priority—play. The level of their irritation made no sense to us until we were forced to come face-to-face with our choice to ignore their clear and unabated warnings.

Something broke.

We'd made a mess tracking dirt, mud and small creatures on freshly swept floors.

We'd left the door wide open.

Harsh, strict rules were enforced. Play was ended. The adults made us stay in the house and clean.

Many of our inactions in response to climate change today mirror the actions of Katie, Daniel, Marcus and I. Now with children of my own, I understand how constantly opening the door, "lettin' all the good air out," was not only costly but affected everyone in the house. The earth is like a house. Regardless of how the house is cooled in summer or heated in winter, if there are cracks and crevices, whatever is outside is going to come in. Once that perfect indoor temperature is reached, any little crack, let alone a whole door opening, can cause change. An open door or window invites the outside in and the house's response is predictable. Doors expand and constrict, foundations shift, paint cracks and walls slant as the house adjusts to the outdoor elements.

We can try to fill the cracks and crevices. No money to fix the window? Tape over it.

Door shifting? Stuff paper or a rag under the door.

This is the same way we've tried "quick fixes" to address climate change. Too much plastic? Tell people to stop using straws. Electricity bill too costly? Unplug your microwave and coffee pot. The solutions are focused on small personal acts versus the larger societal transformations that will be necessary to save the foundation of our planet.

As a climate and environmental advocate with lived experience in Black America, I see climate change as Earth's way of telling humanity to close the damn door or else we're all going to be stuck in the house cleaning the mess we made of our planet. Playtime is over, it's time to clean up.

Climate change can seem as abstract as "lettin' all the good air out" is to children, but lived experience lets us feel and comprehend the effects of climate change. Humanity has responded to climate change not as those who heed the warnings about the consequences of too much play, but as those who are about to be admonished for ignoring the obvious signs.

Climate change is probably an unwise combination of words to influence people. "Climate" and "change" are abstract and subjective. Stuff changes all the time—why is this so important and different? Human activity is largely responsible for the change. Our planet is protected with a carefully crafted, natural barrier of carbon dioxide, methane and water vapors. This balance maintains a temperature, climate and atmosphere that allows life on earth to grow and thrive. It's like one of those delicate, colorful crochet covers your grandma knit with her own hands. It's just enough to keep you warm for a nap but has enough air holes that you can pull the cover over your head and drift into a dreamy slumber without fear of suffocation. But what would happen if someone came in and put one of grandma's big heavy quilts on top of you and the crochet cover? Not only would you burn up from all the body heat trapped in the blankets, your clothes, the pillow, mattress and anything else under the covers would be drenched in sweat. That's what's happening to our planet. Humans are throwing extra heat-trapping blankets on the planet with methane pollution and by burning fossil fuels.

I think of the planet's atmosphere like a Black woman's hair.

Some of us maintain our hair with what's known as a relaxer or perm. It's a choice, a personal preference that does not come without consequence. If you've ever experienced or know a Black person who has a relaxer or perm then you know exactly the physical discomfort and permanently altered state of what it means to heat something higher than tolerable.

Before I began wearing my hair completely natural, I would sit in my beautician's chair every four to six weeks and wait patiently as she applied a white creamy chemical compound to the roots of my hair. The goal was to get the cream close enough to the root to straighten the new hair growth, but not let the cream touch my scalp because it would burn like hell and I'd have to endure flaky-healing chemical burns for weeks. "We'll let this sit so it breaks down good," she would say after applying the white cream to all my roots. I'd sit cross-legged and tell myself the burn wasn't that bad. In reality, I felt like Denzel Washington in the Spike Lee film *Malcolm X*—ready to run to any source of water, including the toilet, just to get it off. Once rinsed out, my naturally soft yet tightly coiled hair was bone straight; it had been chemically altered and quite literally melted straight.

The planet's atmosphere can be visualized similarly. When the planet's atmospheric layer gets warmer, things that weren't supposed to melt actually do melt: Arctic ice, polar ice caps, glaciers. And like my relaxer, we can't immediately reverse it back to its natural state. Polar bears and glaciers may be a long way from my hometown of Greenville, Mississippi, but they are a sign of what we experience from heat trapped gas build up in the atmosphere that impacts us all. Heatwaves get more frequent and more dangerous. Flood risk increases, storms get stronger and wildfires burn more areas. In other words, when it comes to the climate, we are perming the planet's atmosphere and we need to go natural.

Stevie Wonder's 1976 album, *Songs in the Key of Life*, is etched in my mind, one of my favorites. I was not yet a year old when it was released but I faintly recall hearing the song "Isn't She Lovely" played over and over again in our home. Memory tells me it was played by my father to his first-born and only baby girl—me, but it could have been played to and for my mother (I was much cuter, but I digress). One of my favorite Stevie Wonder songs is "As." The lyrics create an impossible order of events that would have to take place before Stevie could ever stop loving his beloved. That's powerful adoration to say that love will last until the trees up and fly away, the ocean covers the mountain and Mother Earth calls it quits. But we live in a time where Stevie's wild declaration is not too far from our reality.

Climate change is not a far-off event. It is our reality and it impacts minority communities first and worst.

At the beginning of 2020, over 14.8 million acres of land burned in Australia alone. In June 2020, the number of fires in the Amazon rainforest rose by 12 percent compared to the same period in 2019—over 2,248 fires were detected. Greenland and Antarctica ice sheets are melting rapidly causing sea level rise that impact America's coastlines. Trees are literally flying away as ash on the wind. The Antarctic ice sheet is bigger than the United States and Mexico combined. Now you may think the Amazon and polar ice caps have nothing to do with your comfy little corner in Houston, Texas, or Baltimore, Maryland, but these ecological systems have kept our air, water and land in balance. When they are destroyed or altered, catastrophic events can occur at a rapid pace.

CLIMATE CHANGE AND BLACK PEOPLE

Climate change is part of the lived experience of Black Americans, whether we know it or not. Familiar sayings perfectly describe the interactions of Black people with nature throughout history. When Grandma starts rubbing her knees and says, "Oh wee! My arthritis is acting up," that means it's about to rain or get cold. "It's hotter than the screen porch to hell" or "It's hot as hell out here!" indicate the temperature is too warm to venture outside in the noontime heat of Mississippi in July.

Colorful adages are familiar masks that hide the stress of survival. These adages also reveal the wisdom tactics that have carried generations of African descended people through unknown and unstable circumstances. But these sayings are also an element of how we embrace and pass along important information—through storytelling, in music and on the stage.

Our relationship to the environment and climate change is similar to our relationship status on Facebook—it's complicated. Many of our Black cultural references point back to various interactions with nature. Sometimes it's a song reference to the protection provided by lakes and streams when our ancestors escaped north from enslavement. Sometimes it's the poetic violence we experienced as "strange fruit" hanging from trees in a breeze. Climate change and environmental justice are so important to Black culture that we verbalize it throughout our storytelling heritage, whether we know it or not. It's a part of our existence.

In the days of a warming planet, global pandemic, quarantine and ever-present systemic racism, I needed to encourage myself that we could and should talk about climate change, particularly what it is doing to Black communities. How does a conversation about climate change supersede a conversation about George Floyd and police brutality or about COVID-19 disparities or voting rights? How do I identify climate and environmental justice issues as relevant, let alone a priority? I connected my own dots between climate change and growing up surrounded by elements of nature in an agrarian society infused with Christian faith.

My formal introduction to environmental justice as a social and community issue didn't come until I was serving as Mayor of Greenville. Lisa P. Jackson, an extraordinary policymaker and President Obama's choice to become the first African American to serve as Administrator of the United States Environmental Protection Agency, made my city her first stop on a tour across the country. Greenville had recently been featured on the front page of the *Washington Post* under the heading "Brown Water," complete with a photo of a Black child in a bathtub full of brown water. The administrator was made aware and contacted us for a visit. Lisa's background is in chemical engineering and, combined with her Louisiana roots, she's a master at constructing solutions that make sense to everyday people. She is smart and funny and within minutes of talking to her you can feel the resilience of someone who's weathered real storms and the warmth of someone who wants to see us all do better. She now serves as the Vice President of Environment, Policy and Social Initiatives at Apple.

"You know this is environmental justice work, right?" Administrator Jackson whispered in my ear as we were standing together preparing to talk to media outside of the wastewater sewer facility in Greenville. I looked puzzled. "No, it's not," I replied. She laughed. "Yes, it is. Infrastructure in a Black community that's dealing with environmental and climate problems is an environmental justice concern." As we continued throughout the morning, I was first perplexed and later angry that no one had connected the dots between poverty and economic disparity in the Delta to environmental justice. As a community that depended on the fruitfulness of the land, developed plans based on the flow of the Mississippi River and maneuvered under seasonal crop dusters, allergies and weather, how in the world were we missing this opportunity?

I went from a little Black girl raised in the Mississippi Delta to mayor of my hometown to a global climate justice advocate. While my overwhelming desire is to help people avoid the consequences of climate change sending us all to hell in a handbasket with gasoline underwear on, I at least intend to pick flowers along the way. Ignoring the climate crisis is like humanity breaking something on the planet while Mother Earth looks on, sighs and says, "See, this is why we can't have nice things."

I like nice things. I want nice things. But nice things shouldn't include the ability to breathe air that doesn't make my children sick and drinking water that doesn't cause cancer. Those are things that should be basic.

MAKE IT MAKE SENSE—CLIMATE CHANGE BASICS

Making climate and environment jargon make sense is the first step to looking at climate problems through a multicultural lens. Together we're going to make it make sense. Throughout the book, I have sprinkled in definitions of climate and environment terms that are closely related to Black culture. I encourage to you think of your own and add them to the list. Here are a few to start:

1.5–2.0 Degrees Celsius

"Every bit of warming matters. Every action matters. Every choice matters." This is how the International Panel on Climate Change (IPCC) says we should think about maintaining the earth's temperature. They've further warned us that the temperature rise is entirely due to human-related behavior. Unless we take action to reduce carbon emissions by mid-century (2050), our families, friends, pets, plants and every living thing that shares this planet will enter an era of irreversible harm. What does that look like? At 1.5 degrees warming, almost 15 percent of the world's population will get extreme heat waves at least once every five years. At 2 degrees warming, that number rises from 15 percent to almost 40 percent. If you think the cost of fish is high now, at 1.5 degrees warming, we lose 1.5 million tons of fish available to catch, let alone sell and eat. At 2 degrees warming, 1.5 million doubles to 3 million. At 1.5 degrees warming, we lose entire species of animals, the sea level rises and the planet warms to the point that almost 20 million people will be displaced.

In the same way that the human body has perfectly and naturally adapted a steady internal temperature to regulate health and optimal function (98.6 degrees Fahrenheit, 37 degrees Celsius is average), the earth has a temperature that is optimal to support life on this planet. When my child's temperature rises from the average of 98.6 to 100.4 degrees (38 degrees Celsius), I know that he has a fever and needs either Tylenol or another remedy to bring his temperature back to normal. As a parent, I look for the signs—I place the back of my hand on his forehead, I look to see if he is acting lethargic, I monitor his appetite and check to see if his hands are clammy or sweaty. If I fail to act and his temperature continues to rise, I am putting his body at risk of organ failure, fatigue or other irreversible harm. A single degree difference can be the distinction between whether or not we can treat the fever at home or if a trip to the hospital is warranted.

All the signs are telling us that Earth has a fever. What's worse is that we are the reason for that fever. How we control, remedy and reduce it is wholly up to us. The good news is that we have climate scientists that are giving triage and doing their best to show us the signs. Dr. Katherine Hayhoe explains, "Over the course of human civilization, the planet's average temperature has been as steady as that of the human body: varying by no more than a few tenths of a degree over thousands of years. Today though, that temperature is changing faster than at any time in our history. The planet is running a fever. And that's why this matters because we are the cause." They are reading the signs that the temperature is rising and it is higher this century than it has been in the ten thousand years that humans have existed on this planet.

Big Green

This term refers to the ten largest and most well-known environmental organizations in the United States. If there were a Divine Nine in the US environmental world, it would be this group. These organizations—Defenders of Wildlife, Environmental Defense Fund, Greenpeace, National Audubon Society, National Resource Defense Council, National Wildlife Federation, The Nature Conservancy, and Sierra Club—are among the oldest in the country and have been touted for their influence and criticized for their lack of diversity in presence and outreach. While the organizations remain mostly white in their membership and leadership, most of the organizations are openly and actively working towards more inclusive solutions to climate change and awareness around environmental justice issues.

Climate Change

Climate changes naturally all the time. But for purposes of understanding the term as a noun—a thing—climate change is the result of human activity causing the planet to warm faster than it should on its own. When we burn coal, gas and oil, cut down and burn trees, we are releasing massive amounts of heat-trapping gases. The worst of these are carbon dioxide and methane. The warmer the planet gets, the more severe weather we experience—heatwaves, stronger hurricanes, bigger floods. Climate change hits minority communities and often impoverished communities first and worst because Black and brown cities and neighborhoods are often geographically located in the least desirable places and are the least protected from extreme weather events due to the historic and systemic lack of infrastructure investment.

Climate Justice

Achieving climate justice means making sure that those who are hit first and worst by climate change are first in line for protection. It is the quest for equity and fairness in how we prepare to adapt and protect people from future climate change or climate impacts. It is an acknowledgment that climate change does not affect everyone equally and we must prepare and provide additional resources for vulnerable and underserved populations so they are not sacrificed for the sake of others. Climate justice is civil rights for climate solutions.

Carbon Dioxide

Carbon dioxide is one of the greenhouse gas emissions that causes global warming. It is like menthol Newport or Lucky Strike cigarettes—the most common and worst for your health and hardest to quit. The "smoke" from carbon dioxide is referred to as emission and the biggest smokers are industries that burn fossil fuels. Carbon dioxide is the bad air left in the house when the door is open and all the good, clean, pollution-free, oxygenated air escapes.

Environmental Justice

Environmental justice is the quest for equity in how we apply and enforce environmental regulations in the United States. Environmental justice is civil rights for the environmental community.

Environmental Protection Agency (EPA)

The EPA is the federal agency charged with oversight of US regulations to protect human health and the environment. The EPA is a rule-making and oversight agency—most of the employees are scientists, economists and attorneys. In the late 1960s industries were running wild with mass pollution to the point that we had lakes on fire and acid rain falling from the sky. The Nixon administration created a set of rules to limit how much pollution could be put into the air. Congress agreed and as a result, the Environmental Protection Agency was created.

Extreme Weather

Climate change amplifies weather events. The intensity, frequency and magnitude increases as the planet warms. Floods, tornadoes, hurricanes, cyclones, wildfires, snowstorms and heat waves already naturally occur but with climate change, they trend stronger, last longer and are more costly for both life and property than in years past. Extreme weather events used to be categorized as an Act of God occurrence by insurance companies. Now, we can go to Facebook or TikTok to catch the latest video of a ten-story wave taking down a building or a house floating down the river as a result of a catastrophic flood.

Frontline Communities

In the same way that men and women serve on the frontlines of a war, frontline communities are the cities, towns and neighborhoods that get hit first and worst by the impacts of climate change. These are the people who are the first to lose their homes from floods and hurricanes, but the last to receive aid and support after the storm.

Frontline communities are usually communities of color—Black, brown and Indigenous people who have lived on and tended to the land but lack adequate resources to protect it. Frontline communities are the first line of defense and sacrifice for the rest of the world. Just like a battle, they take the brunt of the force, allowing those behind the lines to prepare and strategize for future attacks. The frontlines can slow the battle. Whatever the storm, hurricane or flood hits first, the frontlines slow it down from damaging the rest of the country. They take the brunt of the force and once they're gone, the next area becomes the new frontline. Frontline communities can also be fenceline communities if they are located in areas impacted by heavy pollution. Port Author, Texas, and the Manchester neighborhood in Houston, Texas, are examples of both frontline and fenceline communities.

Fenceline Communities

Named for the close proximity of people to pollution, a fenceline community is one that sits along the fence or barrier separating the home from any source, industry or plant that puts out any type of pollution including but not limited to air, water and landfill toxins. Similar to frontline communities, fenceline communities are often communities of color and suffer from systemic racism that has allowed companies to pollute right along someone's back step. People living in these communities cannot walk outside their front door without the view of black smoke pouring from smokestacks or open a window without putrid smells taking over the house, or have to turn up the volume on the TV to cover the sound of machinery running day and night.

Fossil Fuels

Fossil fuels are comprised of exactly what the name implies—fossils. But it's not the myth of decomposed dinosaurs and their dinosaur family friends. It's mostly plants and animals—organic materials that turned to coal, natural gas and oil over time. Technology has allowed us to dig through layers of earth and bring these fuels to the surface where we burn them to create energy. The gasoline we put in our cars, gas used in the stove and electricity to turn on lights has often come from fossil fuels. Here's the catch—once the fossil fuels are used up, there is no way to create more. They are also extremely dirty, create pollution in fenceline communities and worsen the impact of climate change.

Global Warming

The planet's temperature is rising. Earth is getting a fever.

Greenhouse Gas

A Greenhouse gas is a gas that traps heat in the earth's atmosphere. It's referred to as a "heat-trapping gas" because that's what it does—it prevents the earth's heat from escaping into space. Remember when one of your cousins or a sibling farted under the bedsheet then pulled the heavy covers over your head to lock in the smell? Until they let you out from under the heavy covers, you were trapped in a smelly prison. Putting heat-trapping gases like methane and carbon dioxide into the atmosphere is like passing gas but instead of stinking it up, we're heating it up.

The heat-trapping gas adds layers to the earth's planet thereby making the planet warmer. It's like someone who smokes inside the house—you don't see the smoke once it dissipates but over time a dark film appears on the wall almost like an extra layer of paint. Getting the smoke smell out is almost impossible but imagine if it was accompanied by the heat of the cigarette fire itself. Smoking indoors is costly. Why do you think there's an extra charge if you're caught smoking inside a hotel room? Cleaning up pollution is just as costly so let's not do it in the first place.

Carbon dioxide and methane are two of the most commonly mentioned greenhouse gases and have the greatest impact on climate change. Nevertheless, whether it's menthol cigarettes, cigars or weed, the impact to the house is the same and getting rid of the smell takes time and money.

Intergovernmental Panel on Climate Change (IPCC)

If climate change were to have a school board, this would be it. The IPCC is a committee of the world's best climate scientists; its members represent almost every country in the world. The committee is charged with advising the world's leaders of the likely global effects of climate change and suggestions to maintain the earth's delicate balance.

Now that we've reimagined and better understand a few standard definitions, let's take a deeper dive into what to do about it and why. At the end of each chapter is a section called "Before the Streetlights Come On." These are things anyone can do to spur climate action. The biggest climate problems facing the planet are before us and unless we want to all get caught ill-equipped and unprepared, we need action steps to combat it today. So before the streetlights come on and we're out of time to correct climate change, let's get some things in place.

BEFORE THE STREETLIGHTS COME ON

1. Read this book cover to cover then share it with friends.
2. Talk about it. Take any and every social justice issue that concerns you, insert climate change and chances are there will be a connection.
3. Normalize climate conversations every day in every way. Suggest a sermon series at your church, mosque or faith-based group. Post on social media, make a TikTok. Join the climate group at your job or better yet, create one.
4. Share stories from your family that connect you to the climate, nature, the environment and survival. Engage children and young people. You'll be surprised at how many connections to climate change exist within your family history.
5. Vote. Vote for a sustainable, restorative climate policy. Vote for people who support strong climate policies. Vote for voting. Whatever you do, vote.

CHAPTER 3

THE EARTH WILL PASS OUT IF SHE DOESN'T TAKE OFF HER WAIST TRAINER—UNDERSTANDING HEAT-TRAPPING GASES AND GLOBAL WARMING

If you've ever worn, or know someone who's worn, one of those exercise sweat suits or waist trainers, then you know exactly how global warming works. People use these devices to aid in weight loss and burn fat. The shaper itself is made of a heat-trapping fabric, often neoprene or polymer and you just put it on and exercise. The suit increases your body heat which then activates your body to sweat. As you sweat, the moisture you expel is trapped inside the suit, creating more heat and in turn, more moisture. When you take it off, the inside is typically drenched in sweat and you may have lost an inch or a pound due to the loss of water weight, not necessarily fat. Some may argue whether or not this practice is healthy. But hey, if you need to get into a dress for a special event, what's the harm right?

When it comes to global warming and climate change, the earth is wearing a waist trainer in the form of greenhouse gases. There are main gases including water vapor, carbon dioxide, methane, ozone, nitrous oxide and chlorofluorocarbons. The one to watch is methane gas. Once put into the air, methane traps heat in the earth's atmosphere making the planet hotter. Just like the waist trainer, methane creates an immediate reaction that causes the planet to heat up and sweat profusely. We experience the earth's sweat as moisture in the form of melting glaciers, sea level rise and extreme weather. Hurricanes, floods, wildfires and heat waves are results of the earth sweating too much.

2020 marked one of the strongest and longest Atlantic hurricane season on record. Storm after storm battered the Gulf Coast and states including Texas, Louisiana and Mississippi. Before cities and towns had a chance to catch their breath from one storm, the next one was forecast. 2020 was so dynamic that there was a zombie storm. That's right, a storm that blew across the Gulf, traveled across the Southeast United States, went up out and back into the Atlantic Ocean, DIED, only to be reborn in the warmer than usual waters of the Atlantic, turned around and came back through the same path again. That's how gangster the storms were in 2020. While the toughness and fortitude of the people in the Gulf Coast region were not to be doubted, no one should be expected to fight off actual storms like Brad Pitt fought off zombies in the movie *World War Z*. More of these extreme storms will occur as the climate crisis continues to strengthen. If we don't do something now, we will all soon be drowning in the earth's sweat. Earth has to take off the waist trainer or we're all going to pass out.

Climate justice principles and common sense say that the first place to reduce methane and invest in climate-resilient infrastructure is in the communities hit hardest by both pollution and extreme weather. Some climate scientists agree that not only is methane pollution a major factor in global warming and climate change, but if we're going to be effective in fighting climate change then we must deal with methane first.

Compared to carbon dioxide, methane doesn't last long in the atmosphere but it packs a punch while it's there. It heats the planet eighty times more than carbon dioxide. It's like a quick and powerful high, cocaine vs. weed, and in the words of Rick James—"Cocaine is a helluva drug." Reducing methane pollution in the atmosphere can slow global warming by almost 40 percent. Everyone seeking to address Earth's care must, at a minimum, acknowledge the disproportionate impact of global warming to vulnerable, underserved populations and understand that methane is a huge contributing factor.

Methane pollution can be attributed to many sources but the biggest are agriculture, natural gas, oil operations and landfills. Agriculture accounts for the most—30 percent of all methane pollution can be attributed to some form of agricultural practice. Oil refineries and other energy-producing facilities account for another 25 percent and are often the most toxic pollution sites. Refineries are peppered around the country and are primarily located in low-income communities and communities of color. These sites not only contribute to poor air quality for the residents that live next door, they pour millions of tons of methane gas into the atmosphere, further weakening our ability to ward off climate change.

The combination of heat-trapping gases like methane and extreme weather is deadly. In 2020, fenceline communities were already quarantined with children home from school and parents that worked as frontline workers and critical needs staff. Once hurricane season hit, the weather added to the worries of people who were sick from living next door to petrochemical and oil refineries. At the beginning of the pandemic, local residents had to figure out a way to not only sustain themselves economically but to make sure that the community remained aware and prepared for the next storm. From the arrival of Tropical Storm Cristobal on June 7, 2020, through the landfall of Hurricane Sally on the 14th, communities along the Gulf of Mexico petrochemical and industrial corridor experienced five named storms in one summer. Throughout the summer of 2020, neighborhoods were learning how to quarantine at home, manage the essential nature of their work environment while trying their best to breathe under a cloud of pollution and the threat of rising racism targeted at minorities.

When the first storm hit, the air was already black from toxic pollution. Typically, petrochemical facilities have to do what's called a burn-off. A burn-off means that the facility is burning off toxic emissions so that if the storm cuts power to the facility or the operations of the facility are damaged, there is not an explosion that would take out the entire area. These burn-offs are often toxic to breathe. A burn-off also includes a shelter-in-place mandate which requires people to stay inside their homes. Imagine riding out a storm—without electricity, air conditioning or water—under a shelter-in-place mandate due to poison in the air—amid a pandemic—while trying to go to school. It is hard to imagine it once, let alone five times. Nevertheless, under a cloud of pollution, an impending storm and quarantine, Black folks did what we've always done—figured out how to thrive and survive. Think of how many lives, how much time, money and property could be saved if we reduced pollution to begin with?

It's not easy. Weaning humanity off of the fossil fuel energy sources that are causing climate change and methane pollution isn't as simple as shutting down polluting facilities or halting the consumption of meat. Families rely on jobs provided by the oil and gas sector. Hamburgers are a staple in the American diet. Many of our day-to-day activities rely on factors that produce heat-trapping gases. Whether it is putting gas in the car, turning the lights on in your home or throwing steaks on the grill, pollution-causing activities are more common and frequent than we think.

To make matters worse, the oil and gas industry—also known as the fossil fuel folks—have devised tactics to purposefully dupe Black and brown communities into thinking that fossil fuel work is good and necessary for low-income and minority communities. I don't know too many folks willing to trade their low-cost heat bill and the family BBQ for the sake of the planet. But the message becomes worse when the oil and gas industry is pumping billions of dollars into pro-natural gas, campaign ads, messages supporting anti-climate political groups and, believe it or not, local environmental justice and faith-based groups. In minority communities, the repeated message that solar and wind power will raise your electricity bill translates into rich white people wanting poor folks to pay for saving the planet. This is the message that fossil fuel lobbyists pushed in 2021. The American Petroleum Institute (API) spent almost $11,000 a day on Facebook ads alone, targeted at minority populations. The message was simple—vote no against higher energy costs and vote no on an energy tax for the poor. It's almost impossible to convince the average, middle-income white person, let alone Black, brown or any poor person, that rich climate advocates aren't trying to steal their jobs and increase the light bill.

They tried it with the NAACP. In 2014, the Florida NAACP chapter was convinced to slow solar panels on housing because of the energy industry lobby. Then in 2017, a study from the NAACP found that over one million African Americans lived within a mile of an oil and gas operation and were 75 percent more likely to live in fenceline communities. As a result, African Americans were exposed to a higher rate of pollution-related health ailments than the average American. American Petroleum Institute (API) said nope, African Americans have more health ailments simply because they're Black. API money poured into NAACP local chapters and in 2018, the California NAACP chapter opposed government programming that supported renewable energy. The fossil fuel energy lobby machine was pouring millions of dollars into targeting minority-focused organizations around the country and it was working. Every single tactic was designed to create a question of job security or to create doubt in the facts. The NAACP fought back by establishing its own environmental justice division and educating members about environmental advocacy and rights. Led by award-winning environment and climate justice advocate Jacqueline Patterson, the NAACP's *Fossil Fueled Foolery 2.0* report listed the top ten ways the fossil fuel industry works to protect their bottom line versus doing what's good for you and me. It serves as an invaluable reminder that every message to our community should be not only vetted, but shaped and shared through trusted channels that we've cultivated over generations. As my pastor would say, "In God we trust, but all others we verify."

Imagine that person who's worn the waist trainer for so long and has been told they look good because of it. "I mean, it's hard to breathe but I'm getting used to it. It feels ok. It must be all right…right?" The longer it's working, the harder it becomes to take off. The first-world reliance on coal, oil and gas happens under the false impression that there isn't that much harm being done. But that does not mean that harm isn't happening.

Black people—as do all communities of color—have a lot to contribute to the climate conversation. Having our voices in the room will aid in the development of policy, equitable solutions, adaptation and resiliency that both empower and embolden people to do it for themselves and the world. As card-carrying members of the earth's ecological club, equitable climate action and climate justice are the dues we all pay to stay here. But failing to address lopsided and inconsistent impacts means that some people have been carrying the weight of others.

It all starts with remembering our history of resiliency—remembering how we adapted to the barriers of nature in the evolution of our freedom. We don't need waist trainers and sweat suits to be healthy, whole and free. Some of us were built like apples and pears and that is okay. The planet is too. The lessons of adaption and reliance on natural elements and boundaries are born from the stories we share of survival. We must show up for the conversations and see ourselves as part of the climate story.

MAKE IT MAKE SENSE—GLOBAL WARMING

Greenhouse Gases (refresher!)

Gases that behave like sponges. They absorb radiation and heat that contribute to the greenhouse effect of the earth thereby making the planet hotter. There are six main gases: water vapor, carbon dioxide, methane, ozone, nitrous oxide and chlorofluorocarbons.

Methane Pollution

Methane pollution is a byproduct that, once put into the air, traps heat in the earth's atmosphere thereby making the planet hotter. Methane doesn't account for much of the earth's carbon emission, but by reducing this one simple gas, we can slow global warming by 30 percent.

Burn-Off

Burning up the stuff that's dangerous and unwanted. When facing a potentially dangerous storm, chemical industries and refineries use this process to get rid of gases that may cause the entire facility and surrounding area to blow up if the power is disrupted. Burn-offs are considered proper maintenance for facilities. However, when executed, they sometimes place tons of toxic pollution into the air.

Shelter-in-Place

Stay put order. A shelter-in-place mandate usually accompanies a burn-off notice so that local community members are aware of the dangers outside. Problems occur when facilities fail to notify the public and shelter-in-place notices are either too late or don't go out at all.

Renewable Energy

Energy sources that don't run out on us. Solar and wind are considered renewable energy. Somewhere in the world, the sun shines and the wind blows regardless of what we do. Renewable energy can be mass-produced and is affordable.

BEFORE THE STREETLIGHTS COME ON

1. Read the NAACP's *Fossil Fueled Foolery* reports 1 and 2.0. Educate yourself, your friends and family on the costs and benefits of renewable energy as well as the myths. Ask your energy provider for renewable energy options available in your community.

2. Identify your public service commissioner or energy cooperative association and representative. Trust me, you have one. Go to a meeting and listen. Ask about the plans for increasing renewable energy options in your area. Identify any subsides or programs for low-income communities.

3. Do you live in an area close to oil refineries, petrochemical facilities, the natural gas industry, CAFOs (Controlled, Animal Feeding Operations) or a landfill? If the answer is yes, congratulations! You reside in a proverbial

trap house for greenhouse gases (If you don't know what a trap house is, I encourage you to Google it then plan a trip to the Trap Music Museum in Atlanta. You're welcome.). Expose excessive pollution activity by researching emission violations, enforcement rulings or corrective actions from any polluting facility in your area.

4. Make sure there are renewable energy and climate aspects of the science fair at school. Support and promote educational opportunities for kids to explore and learn about global warming and how to fix it in your local area. Their ideas may surprise you!

5. VOTE—for people and policies that reduce methane and other greenhouse gas emissions, build climate resilience AND center equity as part of the solution. Look for local and state goals with targets of reducing greenhouse gas emissions by 2030 or 2040.

CHAPTER 4

LIVING OVER THE SOIL AND UNDER THE CLOUD OF CONTAMINATION—NATURAL GAS, PETROCHEMICAL POLLUTION AND CLIMATE CHANGE

Some people could look at a mud puddle and see an ocean with ships.

—Zora Neale Hurston, *Their Eyes Were Watching God*

I was standing under the carport of Sharon Lavigne's home in St. James Parish Louisiana in June of 2020 when I gained a whole new respect for the term "resilient." Along with a few colleagues from Moms Clean Air Force, I'd made the trek from Mississippi down to Louisiana to speak at the Rise St. James Juneteenth event and put my feet and hands to work raising awareness about stopping the proposed Formosa plant from locating in the community. At the time, there was a glimmer of recovery across the country. The

COVID-19 vaccine had been released and people were getting the shots. President Biden set a goal of having 70 percent of American adults vaccinated by July 4th and my husband and I were fully compliant. Travel was beginning to spring back to life, masks were required on planes and trains. Making it more exciting was the fact that Congress and the president signed legislation making Juneteenth a federally recognized holiday. It seemed right to celebrate the day that the last enslaved Africans got word of freedom by working to make sure their descendants were freed from the petrochemical pollution that encapsulated every aspect of their environment.

I told Sharon I was on my way and ready to do whatever she needed done. Without missing a beat, Sharon led us to her carport that was filled with Rise St. James materials and handed us petitions, flyers and signs.

"Go to the neighborhoods down the street and tell the people that in case of a hurricane, we're gonna meet at Rose's Catering instead of the Formosa plant site."

My mind said, "We're still meeting in the middle of a hurricane?" but my mouth knew better than to question it let alone give it voice. Nothing, including the first named tropical storm of the 2021 season, Claudette, would stand in the way of ending historic environmental injustice in the community. Tropical Storm Claudette would have to wait because we had work to do and justice was impatient. I reached for the extra mask in my pocket and responded, "Yes ma'am, which street do you want us to hit first?"

Sharon Lavigne is the founder of Rise St. James, a faith-based, grassroots organization committed to fighting racial and environmental injustices in St. James Parish, Louisiana. For sixty-nine years, Sharon has been a lifelong resident of the South Louisiana community, also known as part of Louisiana's Cancer Alley. A special education teacher by training, she founded Rise St. James when residents discovered that the State of Louisiana approved yet another petrochemical plant to be permitted right in their front yard. In 2019, Chinese company Formosa Plastics was slated as a 9.4 billion–dollar project that would bring twelve thousand jobs to the region and be located on twenty-four thousand acres of land on the west bank of the Mississippi River in St. James. The community had enough.

Sharon helped organize residents to show up at meetings, sign petitions, write letters and call officials. She talked to newspapers, TV reporters, local, state and federal officials. She reached out to national environmental groups and garnered legal support to file lawsuits. She led efforts to push against permitting applications that allowed construction to begin. In November of 2020, the US Army Corps of Engineers suspended its permit for the proposed Formosa plant and in August of 2021, Rise St. James successfully got the US Army Corps of Engineers to require a full environmental review due to environmental justice concerns.

Sharon's efforts have garnered international attention. She is the 2021 winner of the Goldman Environmental Prize for North America and has appeared in countless magazines and news articles. Her resilience is born from years of weathering storms, regardless of whether or not they are hurricanes, back-to-back floods, relentless layers of petrochemical pollution or plain old southern systemic racism. For years, communities on the frontlines of pollution and climate change have survived through the sheer unwillingness to be deemed the sacrificial lamb for America's dependence on natural gas and fossil fuels. We've had no choice but to figure out ways to raise awareness of environmental injustice while addressing systemic racism, social justice and developing the generational sustainability of the community. Sharon was doing what generations of our families have simply HAD to do—yell until somebody listens.

But less than two months after Juneteenth, I would have another conversation with Sharon as she sat under her carport, but this time under much different circumstances. I called my friend to check on what she needed as Hurricane Ida tore through Louisiana. As the hurricane barreled through the Gulf region, it hit Louisiana hard. Winds knocked out the power and wrapped most of the metal roof around the chimney of her home, she told me, but her resolve and righteous indignation held fast. Once again, the same irresponsible petrochemical plants Sharon has been organizing against for years were releasing toxic pollution under the cover of the storm.

As Sharon waited under the carport for a roofer who was hours late, we talked about resilience. There were gaping holes in her ceiling big enough to shed sunlight and elements on the furniture passed down to her by her mother. We talked about how Black women in the South like us know what it is to be bent, but never broken. About how the recovery from this storm—and the next one—is going to have to honor the power of resilience with a new approach to infrastructure and policies that are just as resolute. Even though the rain was over, the threats weren't. With no power or running water and all of her personal loss, Sharon Lavigne sat under the carport with determination and the full understanding of something deeper. For fenceline communities like St. James, surviving the storm is the easy part. It's the aftermath of heat mixed with the onslaught of millions upon millions of pounds of toxic pollution that will kill you.

Communities like St. James Parish not only take the brunt force of the storm, but carry the burden of recovery under the constant and deadly cloud of petrochemical, oil and gas pollution. Some people think of these communities as "sacrifice zones"—a sacrifice of the lives and property of people who have lived in these areas far longer than any industry or manufacturer was ever permitted to exist in the region. Communities on the frontlines of climate change and pollution, places like St. James Parish, have had no choice but to confront these injustices because disparate environmental harms in the US are not accidental. The racist practice of redlining falsely labeled neighborhoods as "hazardous" in the 1930s and depressed property values. Polluters moved in and funneled clouds of toxic air into the sky, rendered public bodies of water and drink sources unusable and unlivable. Today, the health burdens fall disproportionately on Black,

Indigenous and Latino communities, because they are the most likely to be living where polluters found the least political resistance to operate.

Less than seventy-two hours after Hurricane Ida passed over Louisiana, an analysis of facility records and power outage data showed that at least 138 industrial sites that handle large amounts of hazardous substances in and around parishes that lost power, forcing facilities to rely on precarious backup power systems. Adding insult to injury, the Louisiana Department of Environmental Quality warned that more than a third of its ambient air monitoring sites had stopped working, primarily because of power outages. Within seventy-two hours of the hurricane making landfall, the EPA had received twenty-eight reports of possible spills and pollution events in places hit by Ida, including seventeen possible air pollution incidents. The vast majority of these incidents took place in low-income minority communities that sit on the fenceline of their tormentors.

AMERICA'S LOADING DOCK FOR GAS

Southeast Louisiana is an essential gateway for the nation's oil industry. Almost 20 percent of American oil supply is delivered in the same spot that Hurricane Ida made landfall. Similar to the conditions around retail store loading docks, the backdoor is a constant flux of fast movement to get product inside, without care to the outside surroundings. Oil spills on the concrete in the back are fine as long as the parking lot in the front is clean and inviting. There is no need for greenery and plants in the back. Those resources are placed where customers can see and feel welcomed. As long as the smell of exhaust fumes and toxic materials never wafer to the front, they're fine. The communities along Cancer Alley in Southeast Louisiana are treated as the world's back door by big fossil fuel and petrochemical industries. Oil spills, toxic exhaust from industrial flares and unkempt infrastructure are ignored while companies place emphasis on convincing consumers that they are producing "cleaner" fuels and showing the value of their product as a need our country cannot do without.

America is heavily dependent on this particular loading dock and any disruptions cause massive domino effects throughout the world as well as the immediate needs of the country. Overnight, markets could drop and gas prices increase, thus a priority is often placed on securing the facilities ahead of any storm threat and plans made to quickly get the plants back up and running.

It is important to note that the process of shutting down a plant adds to air pollution. Facilities do something called flaring to release excess toxic gases. Mixed with natural gas and oxygen, it keeps the chemicals from building up to dangerous pressures and emits dangerous and deadly pollutants.

Shell Oil Company's Norco refinery was one of multiple Louisiana oil refineries that shut down the Friday before Hurricane Ida. After Ida passed, black smoke darkened the skies over Norco rising from bright red flares (claimed to be natural gas flaring) and images on social media showed flooding around Norco's parking lot and storage tanks. Norco is situated within Louisiana's Cancer Alley and residents complained that the post-hurricane flaring levels were not normal. Air monitoring at industrial facilities is helpful to monitor conditions that are harmful to human health, but they are not mandatory. Some sites, like the Valero Refinery in St. Bernard Parish, publicly stated that they had shut down their air monitors ahead of the storm to protect the equipment. The priority was placed on product, not people. While the nation trained its eye on rescue and recovery from the hurricane, the equally if not more deadly polluting of people went unnoticed.

The incidents that do get reported paint a picture of uninhabitable environments. Koch Nitrogen in Hahnville reported the release of fifty-eight pounds of ammonia through a flare during a power outage caused by Ida. Shintech Louisiana in Plaquemine reported the release of an unknown amount of ethylene dichloride from a storage tank into the air "due to power consistency/Hurricane Ida." ExxonMobil in Baton Rouge reported releases of nitrogen oxide, nitrate, sulphur dioxide and hydrogen sulfide due to an upset caused by Ida. Mosaic Fertilizer, right down the street from Sharon's home, reported ammonia vapor released inside its St. James facility after a flare blew out during Ida.

In addition to the burden of rebuilding with the inability to breathe, rural minority communities are faced with attempting to rebuild an infrastructure ill-equipped to sustain existing pressures let alone the adaptation necessary to meet the climate crisis. From my experience as mayor of a Mississippi Delta town, I know firsthand the struggles faced by rural and poor communities struck by the climate crisis but lacking the resources and tools to meet the moment. I saw wealthier and whiter areas across the state prioritized because of their higher property values—even as two five-hundred-year floods in eight years battered our community.

The toll taken on small rural communities is both physical and mental. Municipal budgets are ill-equipped to handle the economic strain of storms and the constant wave of weather deteriorates what little tax base exists to support the existing infrastructure. Full recovery was a dream. Through the two floods I witnessed in Greenville, the strain on our municipal budget was immediate. Firefighters and police officers that worked overtime to save lives and property needed to be paid. While they worked, government agencies and large nonprofit organizations were slow to reimburse, and the media quickly forgot about us and moved on to the next tragedy. Federal and state support slowed to a turtle's pace as they had to hit the bigger more populated areas first.

Large donor groups' funding dwindled as their dollars followed the request of their base of donors who were unfamiliar with the needs of our communities. The press and cable news networks were present and on site in the hardest-hit spots, but never stayed to show the harsh reality of recovery for poorer more rural areas. There's no hotel to stay in, no restaurant to eat in, nowhere to plug in a phone. We watched it happen with Hurricane Ida. Within forty-eight hours of images of Hurricane Ida devastation to Grand Isle, Louisiana, the media had moved to the subways of New York, seemingly agonistic to the tough road ahead for those who were charged to rebuild. As Sharon sat under the carport waiting for the roofer to put the tarp on her house, the news had already moved on.

Sharon's tenacity is the embodiment of frontline environmental leaders that for years have refused to take no for an answer while teaching the rest of the world to pay attention to how we forge climate solutions that are durable and maintain faith despite the odds. As we sat on the phone, she said, "I'm highly blessed because no one died. You can always get another home, but you can't get another person. Get me a tarp, and I can do what I need to."

Sharon didn't have to repeat herself. I heard and felt her words, the words she didn't need to say. This resilience is shared, carried through our blood and born from years of weathering the storms of being Black in America. Our parents and grandparents lived through Jim Crow and segregation. We are breaking barriers for access to equal rights today. We do what we need to in order to continue building resilience because real resilience is born of recovery. This is the lesson Black environmental justice activists have been trying to teach through action. On the day I spoke with Sharon, Dr. Beverly Wright of the Deep South Center for Environmental Justice had already reached out to Sharon to make sure she was able to share her story with journalists. By the next morning, MacArthur Fellow and renowned environmental justice activist Catherine Flowers had organized a truck of supplies to arrive at Sharon's door. Both women are highly respected environmental leaders in their own right and they knew the unspoken priority of making sure that Sharon had the resources to continue her work. We come together in this community. It is what any one of us would do to protect home. Having to function in spite of and against the odds has taught us that we must coordinate and work together to make sure the community is not sacrificed.

The refusal to be sacrificed is why we need all levels of governments, political leaders, large environmental groups and concerned people to pay close attention to the strategy. It will take all of us, and we must keep the injustice squarely in our sights, lest we forget. Frontline communities are serving as a shield for the rest of our country and our comfortable way of life.

To build back better is to build back resilient. We have no excuse to give up under the premise of not knowing enough about environmental and climate work or it being too hard, nor can we afford to look the other way as the rampant injustice of polluting people supports our comfortable way of life. If a Black special education teacher in South Louisiana can pull together friends and neighbors to stall a multi-billion international chemical company and keep them from polluting in their front yard, even in the middle of one of the worst hurricanes since Katrina, then any justifications for ignorance of environmental injustices and inaction on the climate crisis is no longer valid. Simple recovery is not enough. Recovery will take time, but justice is impatient and resiliency is an ever-present opportunity. It is resiliency that we must seek and we must seek it with all the gusto and passion of a Sharon Lavigne under her carport after a hurricane, knowing full well that the work could get blown off track again with another storm. Nothing should stop any of us from demanding equity and collectively working to defeat the forces threatening the health and safety of families and communities on the frontlines of the climate crisis.

In the words of Sharon Lavigne, "Not today. We're not going to let it slide and we're gonna keep on working and keep on telling the truth. This storm ain't gonna stop our work."

That's right Sharon. Our work is just getting started.

MAKE IT MAKE SENSE—CONTAMINATION AND POLLUTION TERMINOLOGY

Carbon Capture/Sequestration (CCS)

The process of trapping and storing harmful pollutants before they have a chance to hit the atmosphere thereby reducing our human impact on climate. In other words, catching the smoke before it hits the walls of the house, reduces the property value and costs the homeowners a ridiculous amount of money to clean up. CCS grabs all the bad air so that the good air circulates, keeps the house cool and the family healthy. CCS is disputed in some places. It entails a process akin to catching the smoke in a filtration system inside the house but then forcing it into a plumbing system that would transport the smoke out of the house and into a storage container buried in the backyard. The process may drastically improve the air indoors but who knows what it would do to the backyard. While the technology is here, fenceline and frontline communities have expressed deep concern as to who, how and where the pollutants will be stored since the industries that would be most impactful to capture carbon are not only in their neighborhoods but have been the biggest bad actors in putting pollution out in the first place.

Carbon Tax

A fee paid by one industry to another or a governmental entity for the production of bad air. These are the fees associated with the cap and trade process. Similar to a cigarette tax, it is not paid by everyone, only those who produce it. It has been termed the cost/fee to pollute or polluter payments. Cap and trade can work if the carbon tax paid is sufficient to account for past injustice to the communities most dismantled by the pollution and prevent further harm. That is in ADDITION to adequate and accurate disbursement of funds directly to frontline and fenceline communities. Otherwise, it's like trying to use a coupon for Publix at McDonald's—it won't work. They may both sell water but the coupons for Publix present no value for McDonald's.

Petrochemical industry

The middle-man industry that turns fossil fuel materials like coal and oil into usable items. They turn oil into the natural gas we use to heat homes. They transform natural gas into plastics.

Sacrifice Zone

Areas of the United States where people live in close proximity to polluting sources, with evidence of negative health disparities, economic and educational impacts, yet more and more pollution is allowed to continue. Often, these areas are underserved communities that are on both the fenceline of pollution and the frontline of climate change. They are impacted first and worst by climate change yet there is no real effort to relieve them of the polluting sources that are the causes of said harms. As a result, they are being "sacrificed" for the economic prosperity, health and safety of the country. "Cancer Alley"—the 70 plus mile stretch of petrochemical companies and oil and gas refineries in South Louisiana—is an example of a United States sacrifice zone.

BEFORE THE STREETLIGHTS COME ON

1. Identify all potential sources of toxic chemical releases within your zip code. Start with the Environmental Protection Agency's Toxic Release Inventory (TRI) map to identify any releases that have occurred in the past and to be aware of any steps that have been taken to prevent it in the future. The TRI Map can be found on the EPA website at www.epa.gov in a section entitled, "Where You Live."

2. Items in your home and car emergency weather kit should serve as environmental protections.

Masks are useful in case of toxic air release, as well as gloves and safety glasses. Include battery backup and adapters. When adding clothing to your emergency kit, include items that will protect the skin from exposure to elements and chemicals—long sleeve shirts and pants. Don't forget to cover your head! Add hats to the emergency kit.

3. Encourage young people and or youth-affiliated organizations to develop pollution prevention and extreme weather resources for the community. Children and the elderly are among the most vulnerable to pollution and extreme weather but we can help prepare both through partnership. Girl Scouts, Boy Scouts, Jack and Jill Inc., sororities, fraternities and debutante groups—any civic-oriented group can adopt pollution awareness and extreme weather preparedness as a tenant of one of their programs.

4. Invest in a home generator and look at alternative energy sources as backup power for your home or business. Even if we stop all pollution today, extreme weather events will increase in magnitude and number.

5. Vote for people and policies that operationalize climate adaptation through infrastructure and jobs.

CHAPTER 5

REDLINING BLACK FOLKS IN A GREEN WORLD—FEDERAL DISCRIMINATORY HOUSING POLICY AND ENVIRONMENTAL INJUSTICE

"What are you gonna do different from the other folks?"

Those were the first words I heard from Mr. Jimmy Smith, a seventy-plus-year-old and lifelong resident of the historically Black neighborhood of North Birmingham, Alabama.

I was the newly sworn thirty-nine-year-old regional administrator of the EPA's (Environmental Protection Agency) Southeast region. It was my responsibility to oversee federal environmental protections in eight states, including Alabama and six federally recognized tribes.

He was a Black grandfather on the front lines fighting for environmental justice and air pollution in his own backyard for longer than I had been alive. Mr. Smith had seen the likes of me several times over. To him, I was another newly minted politician from the federal government here to survey the land and make promises I couldn't keep. We were both facing the challenge of tackling years of systemic and racist redline housing policies that allowed pollution to pour into schools, churches and homes across North Birmingham, Alabama.

Housing policies that created neighborhoods like North Birmingham were born of systemically racist housing programs that institutionalized environmental injustice. Environmental vulnerabilities disguised as permitting regulations were written into law and financing policy. For example, the Federal Housing Administration (FHA) refused to insure loans for housing in African American neighborhoods. The FHA's *Underwriting Manual*, first published in 1935, recommended that highways be placed in African American neighborhoods as a separation from white suburban developments. On the white side of the highway, permitting regulations placed commercial and industrial developments—and then green spaces—as a barrier between the highway and housing. On the Black side of the highway, the road was allowed to be built in the front yard of the house. Today, African Americans carry the weight of more exposure to transportation pollution than any other demographic in the United States. Not only does transportation pollution disproportionally impact Black communities, but the known health risks of living close to transportation routes are various including asthma, dementia and premature death. Removing barriers ingrained into the American system of finance and government would be no easy task.

HISTORY OF REDLINING BLACK PEOPLE

In the early 1930s, President Theodore Roosevelt developed a New Deal package that included help to relieve the mortgage foreclosure crisis of the Great Depression era.

But that was for white people.

Although the Roosevelt administration included a "Black Cabinet" and employed more African Americans than any other presidential administration in United States history, the plight of Black Americans during the 1930s wasn't exactly high on the priority list for the federal government. After the Great Depression, the unemployment rate of Black Americans was almost triple that of white Americans. The Great Depression became another milestone of the ongoing "Great Migration" for Black people—over 1.7 million Black Americans moved north in search of jobs, housing and a safer place away from the lynching and violence that accompanied the economic downturn. But the country's solution to improving the economy, managing the population shift and curbing violence was to create more separation and protections for white people.

In 1933, The Home Owners' Loan Act was created and soon thereafter, Congress created a new federal agency called the Home Owners' Loan Corporation (HOLC). Throughout the life of the program, the HOLC controlled more than one in six of the country's mortgages. The HOLC was not a lending agency, but a refinancer of foreclosed or defaulted home loans.

The HOLC also developed a system for rating these loans called "redlining." Redlining was the practice of outlining, in red ink, the neighborhoods that were determined to be too risky for government assistance. Redlined areas often held homes that had higher mortgage interest rates, older homes and polluting industries within close proximity. Another main factor in determining whether an area was outlined in red as hazardous or green as desirable was the race of the residential population.

Despite the financial stability of a Black neighborhood, a white neighborhood was more likely to be deemed green and a Black neighborhood was generally red. Factors like high capacity homes versus single-family homes were discriminatory to the multi-generational family dynamic and history of Black families in America. A house occupied by three generations of a Black family—all United States citizens, gainfully employed, providing multiple streams of income, with one or two older residents staying home to care for the house and children—would be considered less stable and more high-risk than a white family of four with an employed father who has one stream of income and married to a stay-at-home mother of two children.

Black families were not the only ones to suffer from the failure to value the multi-generational cultural dynamic. Similar disparities were experienced in Asian, Latino, and Indigenous communities. However, redlining forms specifically put in writing "Negro" as a metric. Black American families suffered from this direct, prescribed racism, a legacy of the history and systems of enslavement in America.

Housing in the 1930s and 40s was legally segregated and as a result, majority-Black and Latino neighborhoods were almost always redlined. To be redlined on any of the HOLC residential security maps was a death sentence to the growth of a community. Based upon the HOLC assessment, redlined neighborhoods were not afforded the same infrastructure, environmental protection and capital investment as more desirable and majority-white neighborhoods.

Redlining created a barrier to capital investment by banks and other lenders for the purchase of real estate and home improvement. Basic public infrastructure like sidewalks, wastewater sewer systems, waterlines, schools and fire stations were affected by HOLC redlining because public infrastructure is paid and maintained by the tax base where it sits. Moreover, the value derived from the tax base is determined by the property values. If the property values in the area are low, there is not enough money available to provide for the maintenance and upkeep, let alone the establishment of reliable public infrastructure. If the community was within a redlined district, the cost to invest and do business rose exponentially because if property values declined, the tax base to support future investments and businesses declined with it. The HOLC and later the FHA's determination to exclude Black neighborhoods and Black people from federally backed home loans sabotaged the future sustainability of public infrastructure.

Policies for land use followed suit with redlining practices. Exclusionary zoning and suburban covenants prohibited certain industries and commercial establishments from setting up shop in white neighborhoods. In turn, manufacturing and polluting industries such as rubber plants and steel mills were typically located close to working-class housing for ease of access to the labor force, reduced transportation costs as well as the lax permitting and land-use regulations in and around Black neighborhoods. These industries funneled clouds of toxic air into the sky, rendered public bodies of water unusable and, in some cases, made land unlivable. As polluting industries expanded, land-use permits kept them out of areas where white people lived, thereby forcing expansion in and close to redlined districts.

This information was collected through the federally financed mapping process from the 1930s through the 1950s. HOLC created "Residential Security" maps of over one hundred metropolitan cities in the United States. These maps were not only used by the HOLC, but by the Federal Housing Administration, private banks and other lending entities.

One of the cities mapped by the HOLC was Birmingham, Alabama. Identified as "D1" and outlined in red on the 1940 HOLC "Residential Security" map lies the neighborhoods of North Birmingham, Alabama: Fairmont, Collegeville and Harriman Park. The fill-in-the-blank form attached to the map lists spaces to identify the occupation, average income and percentage of foreign-born families. Specifically typed in was a space to identify the percentage of Negro residents that occupied the community. According to the map, the area was home to clerical and factory workers, skilled mechanics and laborers. While mostly Russian, Italian and Greek families made up the 15 percent foreign-born family identification, Negros were another 20 percent of the population. Since a total of 35 percent of the residential population were listed as people of color, the desirability of the area was notated as down despite good schools, churches, parks and ease of employment. The negatives identified were "smoke, dust and obnoxious odors from city, stockyards, slaughter houses, and industrial plants. Heavy traffic. Difficulty of rental collections. Vandalism. Hospital in area."

Years later, these supposed negatives would translate into the justification of sacrifice zones—geographical areas of such poverty, environmental degradation, extreme weather susceptibility and divestment of infrastructure that both the physical land and people therein are considered not suited for any good use.

We can still see the impacts of redlining on pollution today. Formerly redlined communities are less likely to benefit from enforcement of pollution violations by industrial facilities that sit at their back door. Consent decrees meant to improve conditions in polluted communities rarely provide financial benefit to the impacted community itself, and rarely call upon the polluting entity to do more than corrective repairs and promise to report future mistakes.

FIRST MEETING IN NORTH BIRMINGHAM, ALABAMA

After redlining in the 1930s and 1940s, for the next seventy years the North Birmingham neighborhood continued to decline, and its residents disproportionately suffered increased health disparities and poverty. When I arrived in 2014, I had been warned that the community residents may not be nice and welcoming to my visit. As we planned and prepared for what was to be my first official visit as an EPA regional administrator, I sat through briefing after briefing and listened to the well-put-together plans of my staff. They walked me through the environmental problem as it had been studied—what we knew about the complaints, substances found at homes, PM2.5 counts, plans that had been prepared for removal of contaminated soil and the community engagement the staff had been a part of for the past few years. I was prepared to speak with not just community members but members of the city council, the mayor, the city administrators and coalitions that were dedicated to eradicating environmental injustice and improving air quality throughout North Birmingham.

The plan was to ride through the community before the meeting and get a sense of the layout of the neighborhood. We all boarded black SUVs and other vehicles that were clearly government and proceeded to visit the neighborhood. When we turned the corner, I felt like I was home. "Make a left at this railroad track. At the stop sign please," I said to the bus driver.

The driver looked back in disbelief and responded, "Madame Administrator, if we just keep going straight…"

"Make a left at the stop sign," I quickly replied.

The driver made the left and we came to a corner where I could see an array of houses lining the street in front of me. Some of the houses were shotgun-style—an all-too-common reminder of the plantation and later sharecropping system commensurate with Black families living in low-income neighborhoods in the South. Others were brick homes with wrought iron fences and garden decorations neatly lining the sidewalk to the house. All of them were quiet. As we approached, I saw the questioning looks that people in any neighborhood have when they see a long line of black government SUVs rolling into their neighborhood. Men and women began to appear on front porches. *What's going on? Is that the police? Did something happen? Is somebody in trouble?*

I got out of the vehicle. I was wearing an EPA jacket with the three identifiable government letters on my chest and back. I went to the first house where I saw someone standing outside on the front porch.

I smiled and said, "Hey, how y'all doing!"

"We just fine. Whatcha doin' over here?"

I replied, "My name is Heather McTeer Toney. I'm the new regional administrator for the EPA and I just wanted to come see for myself what's been going on with the dust that I hear is coming from that plant over there."

Slowly but surely, people relaxed as we went from house to house. I sat on front porches with folks who looked like my friends, family and church members from home. One woman came outside and started talking, but once she realized photographers were taking pictures, she ran back in saying, "Let me go take my rollers out of my head."

The hour or so that we spent sitting on front porches was more valuable than any briefing. I was listening and hearing firsthand from Black people and families who had lived generation after generation under a cloud of pollution. I heard their stories from their own mouths, in their own way, in their own vernacular. We spoke the same language—people. I was able to deeply feel not only the concern that they had for their health, but their concerns for their children's health, for their neighborhood and for their economy. The experience allowed me to hear their dreams for a better future. They knew the places where they worked were ultimately making them sick. The place where their family members got their paycheck and later their pension was potentially the cause of those same family members dying of cancer. The company that had provided food for the table and funds to put children through schools, polluted the air they breathed, leeched toxins in to their water and contaminated their soil.

We made our way to the 35th Avenue Superfund site. A Superfund site is an area or location where we've put so much toxic stuff that unless it's completely restored, nothing else can go there. It's like the junk drawer in your house. Over years it's held everything from nails and old screws to ketchup packets and napkins from the drive-thru. Unless you clean out the entire drawer, you don't feel comfortable putting anything there other than junk. Superfund is a nickname. The real name for the legislation that appropriates money for major waste site cleanup is CERCLA—Comprehensive Environmental Response, Compensation, and Liability Act. Established by Congress in 1980, CERCLA ensures that where there is a need for a big environmental cleanup, funds are available but taxpayers don't bear the brunt of the cost without accountability to the state and polluting entity. It can get a bit controversial. If the federal government says that a site is so polluted that they need to step in and provide CERCLA or Superfund dollars, then the state where the site is located has to put up dollars as well.

We were actively working to remove contaminated soil from the 35th Avenue project in North Birmingham while noting other areas where we needed to do more work. The CERCLA money would be a big help but in order to secure CERCLA funds in the amount necessary to completely revitalize North Birmingham, it would have to be listed on the National Priorities List (NPL). Believe it or not, I figured that would be the easy part. One of the ways toxic site cleanup is funded is to assess a portion of the cost to the polluter. If it can be documented that the source of pollution is a specific site, industry or bad actor, the state or federal government has the option to begin cleanup and assess them the cost, whether the polluter agrees or not. I sought the NPL, which required some state funding. Convincing the administration should have been a cakewalk compared to convincing a Black Alabama neighborhood that we could effectively reverse engineer racism. That would most definitely be the hard part.

When Mr. Smith and I met at that first meeting, I immediately felt the weight of his expectations. Slight but strong and upright, the tendons in his dark chestnut arms accompanied by the neat gray of his short afro and beard let me know he meant business. But I also sensed an element of sadness behind the apprehension in his eyes. Here was a man who had suffered great pain as a result of environmental injustice. He carried with him photos of family members he'd lost to cancer. What would I do, what could I do, so differently that it would reverse a century of racist federal and state practices? How could I even begin correcting all the environmental injustices and pave the way for economic opportunity and clean air? Who in the hell did I think I was? Jesus's little sister?

Mr. Smith's disposition toward me upon my first visit as Regional Administrator to North Birmingham was not malice, but instead a lifetime of pain, neglect and frustration. We, the federal government, needed an apt understanding of the redlining and housing discrimination history behind pollution in North Birmingham and our role in it. Without an accurate and sincere acknowledgment of the history, preparing to present solutions at community meetings was an exercise of futility. He stared at me, waiting to see if my response to his challenge "What are you gonna do different from the other folks?" would indicate actual knowledge and appreciation for the struggle of North Birmingham residents and quite frankly, poor Black people everywhere. His tone, polite yet firm, indicated I might be just another politician that had only read about Black communities and would be gone before the streetlights came on. It was up to me to prove otherwise and I fully intended to.

I have a theory about community meetings. It takes three meetings to get started on the real work, but it's the first meeting that sets the tone. The first meeting is an information gathering, who's who of all the players. In the first meeting, you'll find controversy and confrontation. People vent. People are upset. People have questions. During my first meeting at North Birmingham, all that occurred and more. Mr. Smith came to the front, took the microphone and with pictures of his deceased family members in hand, asked me what I was going to do differently. After the meeting, he and other residents warned me about the executives of the company. I was told that despite efforts to clean up, somebody at the state or federal level, either the governor or the head of the Alabama Department of Environmental Management (ADEM), always halted the process. At that first meeting, I promised that as long as I was in the position of regional administrator of the Environmental Protection Agency Region 4, I would do my best to follow through with cleanup and communicate honestly about the process. I committed to doing as much as we could in the time we had.

Pollution is like wind—you may not always see it, but you can feel it and know it's there. No person or thing escapes pollution when it's present in a community. Corner grocery store awnings are covered in the same particulate matter dust that settles on cars sitting in church parking lots. There is a slight shade difference on the playground slide after a child slides down, denoting the line of dust film. Sit on the front porch long enough and you can feel the tingling sensation that something is constantly crawling on your arm, despite steadily swiping and wiping at something you cannot see. What's scarier is knowing that the same particles are going in and out of your lungs just by breathing. This was the lived experience of North Birmingham residents every day.

The biggest alleged offender was the area coke plant, Drummond ABC Coke, Inc. This isn't your bottle of Coke, Diet Coke or Coke Zero sitting in containers ready to be shipped to the nearest grocery market. The coke produced at these facilities is purified coal. It is widely used in the steel industry. The process for making coke requires heating coal to extremely high temperatures to remove impurities, then cooling it with water. As a result, huge plumes of steam burn off into the atmosphere creating harmful greenhouse gas emissions and carcinogens—particles known to cause cancer.

Despite my youthful, naïve, we-can-do-anything-but-fail attitude, I would soon come to realize that Mr. Smith was right to be skeptical. I was not the first regional administrator attempting to make significant corrections to the legacy soil contamination in North Birmingham. Before I resigned from my role as regional administrator, as was practice with every outgoing administration, I'd be denied requests to place North Birmingham on the National Priorities Listing to receive additional cleanup money, be criticized by high-ranking state officials and even taped on a wiretap. I learned just how far industry leaders and some state elected officials would go to protect polluters and avoid cleaning up this historic Black neighborhood.

But there was something familiar about North Birmingham that reminded me of my own hometown of Greenville, Mississippi. Collegeville, Fairmont, Harriman Park—three quaint, North Birmingham Black neighborhoods were just like Nelson Street in Greenville. The foundation and outlines of a successful working-class family community were evident everywhere, from the antique script on glass storefronts to the neatly kept lawn of the local church. I could attend Hopewell Missionary Baptist Church in North Birmingham, close my eyes and feel the same rhythm and warmth of New Hope First Baptist Missionary Church back home. The structural foundations of single-family housing and schools were apparent despite disrepair, obvious impacts of economic downturn and youth flight. These were homeowners, but surrounded on all sides by pollution in the air, wastewater and soil. You have to cross railroad tracks to get into the community and drive in one way and out the same way. Look to your right and you see a major manufacturing plant. Look to your left and you see the backyards and clotheslines of families that have celebrated life in these dwellings for generations. To me, this wasn't just another Black neighborhood in the South. This was home.

I was determined to make a difference no matter how big or small. I wanted to make sure that I didn't just send staff to North Birmingham to address the community but to personally attend to it myself. I had heard the rumors of sickness and disease. Chronic healthcare problems such as cancer had plagued the community for years. I heard the stories about the dust layered on clothes and cars and how a quick swipe of the hand across any surface would come away with a palm full of an unknown substance that no living thing should breathe in. I'd heard reports that people would go to work at the local coke plant, surrounded by the polluting dust and then come home only to find their yards, their kids' bikes and toys, and their clothing completely covered. Days missed from work, unaffordable visits to a hospital and unavoidable acceptance of disability assistance were common. Many suspected that the pollution was the source of their health problems.

EXCHANGING THE ROSE-COLORED GLASSES FOR ALOE-GREEN GLOVES

At the end of that first meeting, Mr. Smith said to me, "You know Madame Administrator, when you came here today, they were ready. They sprayed the streets with water to clear all the dust. They made sure that we didn't have any type of burn-off incident or visible flare. When you came, they wanted you to think it wasn't as bad as we were saying it has been. But I want you to stop by here one day with no staff. I want you to come through one day unannounced. On your trip home to Mississippi one evening or morning, just ride through North Birmingham and see if what we're saying is true. See how the railcars just stop on that one entrance in and out of our neighborhood and prevent ambulances and police and firefighters from getting to us in time. Come through and see for yourself the dusty layers on our clothes and cars. Come through unexpectedly and see for yourself if what we're saying is true. Then come back and let us know what you're going to do differently."

I took him up on that challenge and on one of my trips from Mississippi back to my office in Atlanta, I drove through North Birmingham. Mr. Smith was right. I sat shaking my head at the railcar that stood dead still on a street, blocking the entrance and exit of the community. I found someone who lived in the area and had them take me into the community through a back way. I saw firsthand the dust they talked about on the streets, in the yards and on the vehicles. I even went by Mr. Smith's house, to stop by and say hello, but no one answered the door. It didn't take long for word to get back to him that a strange car pulled up to his house, and it looked like that lady from the EPA had knocked on his door.

Our EPA team began addressing the concerns raised at that first community meeting while recognizing we were facing systemic barriers that would require transformational change. I had to grow up and exchange my rose-tinted glasses for a pair of healing, aloe-tinted green work gloves. It was time to put hands in the dirt and work.

We identified internal and external sources to leverage what little funding we had to clean up the environmental waste in the community. We asked local community groups to sit with us in the planning and help us to understand deeply the problems that existed alongside the soil contamination that plagued the neighborhood. We looked at schools that had been closed due to environmental concerns as well as population loss and redistricting. They were important because community members valued and prioritized educational restoration and wanted to know how we could repurpose the buildings as part of the environmental cleanup plan. We examined ways that we might revisit economic development and revitalization as cleanup moved forward. We took a 360° approach. The goal was not to come in, clean up and leave, but to strategize alongside the community and identify scalable solutions that achieved environmental and economic wellbeing. I wanted to help plant seeds of real change so that those in the neighborhood could envision success. That meant not just an environmental win but furthering the future legacy of their community.

Within a few months, we garnered the trust of the community and started working together to expand our reach. A strong community revitalization plan had already been developed by the North Birmingham Community Coalition—including housing, commercial development and health. The EPA held summer environmental education camps for neighborhood youth. We conducted a bus tour of the neighborhood bringing, along with our staff from EPA, teams from the Housing and Urban Development (HUD), the Department of Transportation, and members of Congress who represented the area. Mayor William Bell, Councilman William Parker, members of the Jefferson County Health Department, and nationally known celebrities Gizelle Bryant and Erika Liles were part of the coalition. One of my most reliable federal agency partners was HUD. At the time, former Florida State Legislator Ed Jennings Jr. served as regional administrator for HUD in region 4. We covered the same states and were often agency colleagues on the same projects. Every time I said, "Ed, I'm headed to North Birmingham, I need your help," he responded by showing up in person with suggestions, resources and an undying love for University of Florida football, a brave statement in Alabama football territory. All of the collaboration was support to show the community that we were serious about efforts to correct environmental injustices and make the neighborhood sustainable and resilient. We talked together about how we could leverage combined resources for the benefit of the community.

I was not prepared for what came next.

WIRETAPPED? IS IT THAT SERIOUS?

As work progressed, I received cease-and-desist letters on official letterhead from the Alabama Department of Environmental Management, the Alabama governor, Senator Jeff Sessions and a few state Supreme Court judges. All the letters stated that we did not have the right or permission to proceed to clean up the North Birmingham community. I had no idea how far and intense the opposition would go, but I quickly realized I faced an entrenched and wholesale effort across the entire state of Alabama to ensure that the polluting companies had more influence than the people. As we sought to include North Birmingham on the National Priorities List, those who knew that this would require the state to pony up the money to help clean a Black community made it clear the state of Alabama was unwilling to commit funds to do so.

I pushed back. As time drew near for me to go on maternity leave to give birth to my first son, the effort to halt the cleanup in North Birmingham grew fiercer. Later, as we came to the end of our EPA resources and federal authority in the region, I discovered more opposing efforts afoot than I knew. In 2018, former State Legislator Oliver L. Robinson Jr., a Black state representative for a district covering part of North Birmingham, was convicted of accepting bribes from a law firm in Birmingham and a coal company. In exchange for the money, he was to advocate for opposition to the EPA actions in North Birmingham and undermine the process of protecting this community. Unbeknownst to me, I was a target of Rep. Robinson and had been recorded in a meeting with the attempt to find something that would show that I, too, was undermining this community. Instead, those listening to the wiretap got an earful about how elected officials were not acting in the best interest of their citizens. Let's just say that my words would be deemed inappropriate and unladylike by my pastor and mother, but would have made my daddy proud.

I then understood what Mr. Smith had been trying to tell me: regardless of all my efforts and those who had come before me, the state of Alabama, the old antebellum South, a multimillion-dollar company and white state leaders were not going to side with a poor Black community. They would go through hell or high water to protect polluters before they protected Black people. Even the elected official that was supposed to protect the interests of the people could not protect them against the corporate interests. Money talks.

I believed that if I was willing to put myself, as regional director of the EPA and a fellow southerner, born and raised in a Black community not too dissimilar from North Birmingham, on the line alongside community activists like Mr. Jimmy Smith and others, then certainly our collective little Davids could take out the Goliath. Above all, we had faith. If God be for us, then who could be against us?

We are all still fighting to this day and how the story of North Birmingham will end has yet to be written. In 2018, the regional administrator that replaced me, Trey Glenn, was indicted by a federal grand jury on charges of violation of state ethics law roughly halfway into his tenure. For three years the seat was filled by capable career staff but it wasn't until 2021 under the Biden administration that another presidential appointee was selected and seated in the position. Shortly before Thanksgiving 2021, Administrator Michael Regan announced the appointment of Daniel Blackman as the new regional administrator covering North Birmingham as well as the seven other states of the region. In December 2021, the EPA announced new additions to the NPL cleanup list. Alabama has twelve sites on the active list and as of this publication, North Birmingham is not included. North Birmingham remains on the proposed list as it has since we placed it there on September 22, 2014. Mr. Smith lives close to Birmingham but is no longer in the neighborhood. For health reasons, he's moved with his children and longs to see North Birmingham revitalized. When I spoke with him, he shared that he'd recently lost his wife. She is now among the ancestors pushing us to finish this fight. He asked about my family and made me promise to visit North Birmingham, the people and the work yet to be done. There was no hesitation in my reply, "Yes sir." He said, "Madame Administrator, you're doing the Lord's work. Don't you forget that and don't you forget us. You promise me that. This is the Lord's work and he's called you. We are proud of you."

The story is a common one told in Black and brown neighborhoods across this country. Addressing environmental injustice means addressing racism in every shape, form and fashion—including housing discrimination.

MAKE IT MAKE SENSE—REDLINING AND ENVIRONMENTAL JUSTICE

Federal Housing Administration (FHA)

Mortgage lenders run by the United States government. FHA is part of the Housing and Urban Development (HUD) agency. FHA has three secondary markets or private government-backed programs you probably know well—Fannie Mae, Freddie Mac and Ginnie Mae. Fannie and Freddie are the better known of the three, Ginnie is the lesser-known sibling. Believe it or not, they are not named after real people and instead, are nicknames for government programs. Fannie Mae is the nickname derived from the acronym for the Federal National Mortgage Association (FNMA). Likewise, Freddie Mac is the nickname for the program Federal Home Loan Mortgage Corporation (FHLMC). Fannie Mae is the oldest, born in 1938, and the program most affiliated with prohibiting Black people from access to federally backed home loans.

Transportation Pollution

Transportation pollution is a toxic mix of the gas or diesel fuel we use in our car and the fact that it lingers in the same area for hours on end. People living and working in these areas breathe it in, which harms the lungs and overall health. This creates poor air quality and inhibits the health of the impacted region. Transportation pollution exists in high numbers around major highways and shipping channels across the United States. African Americans are more likely to live in close proximity to transportation pollution than any other demographic in the United States.

The Black Cabinet

Also known as the Black Brain Trust. A group of forty-five men and women who served as senior advisors to the Roosevelt administration during his term. None were officially appointed as Cabinet members to the Roosevelt administration but they did represent the largest contingency of Black advisors to a US president at that time.

Home Owners' Loan Act/Home Owners' Loan Corporation

The Congressional action that gave authority to the federal government to enter the home refinancing business. The HOLA was passed in 1933 and created the Home Owners' Loan Corporation. The HOLC was the creator of residential security maps, which facilitated the practice of redlining.

Residential Security Maps

140 city maps created by the Home Owners' Loan Corporation that grade city development potential by color grade. Grade A zones were colored green and considered the best for future development. Grade B zones were colored blue, grade C zones were colored yellow and grade D zones were colored red. The lower the grade, the lower the likelihood of federal or private financing for housing development in that zone. The term redlining is derived from the association of grade D zones with prohibitions against funding minority communities.

Redlining

The practice of qualifying minority, low-income and undesirable neighborhoods as grade D on residential security maps created and used by the Home Owners' Loan Corporation. The practice allowed legal prohibition against financing in minority communities across the United States and was used by banks and real estate developers.

Shotgun House

A dwelling place of formerly enslaved Africans on southern plantations and later, for sharecroppers. The name was derived from the fact that one could shoot a shotgun through the front door and it would go straight through the house, hitting nothing, and out the back door. Shotgun houses were small (one room wide, two to four rooms deep) poorly constructed structures commonly found throughout the South well into the era of sharecropping.

CERCLA—Comprehensive Environmental Response, Compensation, and Liability Act of 1980

The Congressional act that funds the cost of environmental cleanups, especially when hazardous waste is involved and no particular party can be clearly identified. CERCLA is also known as Superfund and the physical location of the cleanup is called a Superfund site.

Superfund

Nickname for CERCLA. The site is massive or there exists extreme waste that it will take an amount above and beyond what any local or state budget could be expected to contribute at one time to be cleaned properly.

Consent Decree

A settlement or agreement between two parties that requires an admission of guilt or liability. This practice is common for environmental cleanups where a business or industry is a party to environmental harm or failure but disputes fault. A consent decree is like a deed in lieu of a house—the homeowner and the lender agree that you turn over the house and they release the lien. You avoid a foreclosure on your credit report and the lender gets the house.

On a consent decree, instead of giving up the house, the business or industry is required to clean up without admitting that they made the mess and it's their responsibility. They are required to pay for the cost of cleanup and attorney's fees. Consent decrees are powerful tools of enforcement used by the US EPA (Environmental Protection Agency) and the US Department of Justice. They can be imposed on any entity that is found to cause environmental harm including but not limited to oil and gas manufacturers, city water systems and local dry cleaners or paint shops. There are disputes about the effectiveness of consent decrees and who they are meant to protect. Decrees can often require cleanup of a specific area of a site, but do nothing to rectify past harms experienced by those who have suffered the exposure. In addition, many of the large companies have more than enough cash to pay the pittance of fee required and the fees paid as part of the penalty of the decree rarely make it to the impacted community.

Coke Plant

Not the place where they produce the dark delicious carbonated substance I consume a bit too much. Instead, coke in this context is the byproduct of coal. Coal is a fossil fuel. It's mined from the ground and then used to create energy, but in its original state it has a lot of impurities. It's worth noting that the charcoal we use in our backyard grills is not the same as coal. Charcoal is a wood byproduct and burns much cleaner than coal. So step away from your uncle's grill—he's not wholesale killing the planet with his signature ribs and chicken.

Coke is what you get when you heat coal to remove the impurities. The coke is then used as fuel, often in steel plants, to create a stronger, more refined product. But it's highly polluting, especially to the communities in which it sits. Parts of the process includes open-air exposure and transport of product by rail lines that run adjacent to neighborhoods and family homes. It's funny—we've long lived with the myth that if you burn coal hot enough, you can make diamonds. Turns out, the thing you make when you heat coal is purified pollution.

Particulate Matter

Itty bitty, microscopic pieces of dirt, dust and bacteria that can cause havoc to human lungs when breathed in. Particulate matter is what makes up air pollution and it comes in many forms including smoke, pollens, soot, smog and viruses. Some particulate matters are considered carcinogens or cancer causers. Particulate matter is measured in size. The smaller the number, the smaller the size of the particulate matter. For example, PM2.5 is smaller than PM10.

NPL (National Priorities List)

The list of all Superfund sites in the United States of America. Placement on the National Priorities List means access to funds from CERCLA or Superfund. In 2021 there were over 1,300 superfund sites on the National Priorities List with an average cleanup cost of $25 to $30 million per site.

BEFORE THE STREETLIGHTS COME ON

1. Look at the list of redlined city maps located in the appendix and determine if your city was a redlined community. Ask your local and state leaders what (if any) changes have been made to ensure compliance with federal fair housing practices. If your city is smaller or not on the redline list, do local research to determine if those same guidelines existed and what changes have been made.

2. If you live in an area or subdivision subject to housing covenants or homeowners associations, check the governing document of the organization or group. When was it last updated? What accommodations are considered to ensure equitable and future sustainable reliance on renewable resources? For example, are you on well water or city water? What is the energy source?

3. Identify and get to know members of your local housing authority and planning commission. This information can be found on your city or town's website. These are typically appointed positions. Consider serving.

4. Prepare playgrounds and plan to play. Ask to see the plan for future housing and development in your community. Look for resiliency

and adaption plans for climate impacts as well as public housing plans. Does public housing have the same access to benefits as the rest of the community, green spaces, parks, sidewalks, provisions for updating wastewater/sewer in structure, etc.? What changes need to take place to provide healthy places for children to play?

5. Vote for people and policies that support environmental and climate justice as a foundational element of fair housing practices.

CHAPTER 6

GRANDMA'S HOUSE IS NOT FOR SALE FOR THE SAKE OF THE PLANET—CLIMATE SOLUTIONS DISGUISED AS GREEN GENTRIFICATION AND BLUE LINING

FROM REDLINES TO BLUE LINES

The process of redlining gave way in the early 1970s to urban renewal—a strategy to develop interstate eight-lane highways, super malls and residential developments that replaced Black and brown neighborhoods, dislodged families, cultures and businesses. Urban renewal took big parcels of public land and private property for what was said to be the modernization and improvement of infrastructure throughout the country. Unfortunately, it meant devaluing minority neighborhoods while destroying historic and often meaningful neighborhoods and landmarks in cities. Federal programs were used to displace entire Black and brown neighborhoods that were stable and a viable part of the culture. As a result, major highways split historically

African American neighborhoods like Treme in New Orleans, Louisiana, Greenwood in Tulsa, Oklahoma, Compton in Los Angeles, California, and Paradise Valley in Detroit, Michigan.

Urban renewal opened the door to rezoning, giving local boards and commissions permission to create de facto barriers around white neighborhoods. Rezoning protected the infrastructure, growth and capital investment of white communities and forced Black and brown communities to live next to pollution and contamination. Today, there are new methods of doing the same thing, and climate change is being used as the excuse to further exasperate the disparities while it should be a reason to correct the past injustice.

Green gentrification is the planned process of creating environmentally friendly green spaces, at the cost of excluding and displacing poor and minority neighborhoods under the guise of protecting the planet. Cleaning up polluted neighborhoods and sites improves property values. To adapt to the impact of climate change and provide protection from storms thus creating more resilient cities and towns, sustainable infrastructure must be developed. But the green gentrification process means that climate change is often used as an excuse to move out traditionally underserved and minority families who are unable to afford the "new" green amenities and housing costs. This creates both a distrust for climate change initiatives and a dislike for those employing the methods despite their best intentions. It's tone-deaf to the intersecting issues faced by those who live in communities most vulnerable to climate change.

Within green gentrification, another practice can be commonly identified: Blue lining. Akin to redlining, blue lining is the practice of banks and lending institutions physically drawing a blue line to indicate water around flood-prone minority communities. The lenders purposefully withhold capital from minorities living in flood-prone areas, while financially supporting the wealthy who can afford high insurance prices.

While loans are not always directly denied, the cost to rebuild to local standards (often controlled by permitting boards and commissions with direct ties to the industry) prices minority communities out. Wealthier people who can afford additional insurance and compliance with newer, more resilient standards have the resources to purchase the property and for redevelopment. Bahama Village in Key West, Florida is an example. Known for tourism, beautiful white beach sands and a lively Duval street crowd, Key West is a popular and valuable area to own property—despite its highly susceptible position to climate and environmental erosion. The historically Black and Indigenous community of Bahama Village has been subject to blue lining through rising housing costs and systemic racism. Community members are faced with the difficult decision to maintain generationally passed down properties under escalating costs or sell to wealthy builders and suffer displacement of home.

Now that we understand green gentrification and blue lining, let me tell you about the time I met Brad Pitt in a hallway in New York.

In September of 2009, I rushed to attend the Clinton Global Initiative conference at the last minute. The big issue of the day was water—how to make sure it was accessible and affordable. I was attending the annual Congressional Black Caucus event in Washington, DC, when Steve Pruitt, our city's lobbyist and later my mentor, called and asked if I could get to New York. Steve was a quiet gentle giant kind of man and an expert when it came to Washington, DC. He served in the Clinton administration in the Office of Budget and Management and he knew EVERYBODY. I quickly came to understand that while he was working as the city of Greenville's lobbyist in DC, he was teaching me how to make my own connections empowering me to build from the seeds he was planting. Steve could do a full day of seminars and panels, take a break, then be ready at the car for six gala evening receptions in full evening wear and an early a.m. diner breakfast to cap it all off somewhere around 3 a.m. the next day. Despite being dog tired from all the CBC functions, when Steve said catch a train to New York to the Clinton Global Initiative because there was a shot at Bill Clinton's foundation being interested in helping to fund water solutions for Greenville, Mississippi, I took a deep breath, pulled out a suit and booked the ticket praying that my credit card would go through. I had to pay for it personally because it wasn't a pre-approved city travel expenditure. I skipped the hotel cost for the night and made it a roundtrip one day affair. No sign of pity for me at all, Steve said, "I'll meet you at the train station." He was thirty years my senior so I needed to figure out how to keep up.

Early the next morning, after arriving and clearing security for the Clinton Global Initiative, Steve and I proceeded to meet and greet a host of celebrities, philanthropists and friends of Steve. Everyone was talking "green" this and "clean" that. I saw Jessica Alba standing on the opposite side of the room. Sitting next to me was Francine Drescher, the beloved character from the sitcom, *The Nanny*. In 2009, green was becoming the new "black." The plan was to be in the right place at the right time for me to meet Bill Clinton and tell him about our project to make sure we have clean, reliable water in Greenville. As the former president of the United States and, more importantly, governor of our neighboring state Arkansas, he was familiar with access to clean water in the ArkLaMiss Delta. Steve said, "He's excited to meet you and he's smart as hell. Tell him exactly what you need and why you want this audience to know about Greenville."

As we stood in a backstage hallway waiting for Bill Clinton to walk by, out came Brad Pitt. Now I've been a Brad Pitt fan since *Interview with the Vampire* so keeping my composure was not easy. Steve gave me that "don't-embarrass-me-child, focus" look. Brad Pitt was preparing to go on and talk about the work he was doing in New Orleans post-Hurricane Katrina with the Make It Right Foundation and was highlighting the economic advantages of green initiatives. I was just excited to say that I got to meet Brad Pitt. President Clinton quickly walked in, grabbed my hand and said, "Mayor, it's such a pleasure to meet you, tell me what y'all got going on in Greenville? Is Doe's still there?" I proceeded to give my best three-minute speech on why the entire world needs to understand that fixing water access in Greenville was a practice in scalable solutions for low-income communities around the country. After he thanked me, Steve and I went to our seats.

Two things happened next:

1. President Bill Clinton walked on stage, gave an introduction to the audience and (unbeknownst to me) proceeded to invite me to come up and tell the audience about Greenville. (I must have given him a helluva three-minute talk!)

2. Photos of the housing Brad Pitt's foundation was putting up flashed across the screen. The first thing that drew my attention was that some of the roofs were flat. I thought to myself, "My, that's an interesting concept for New Orleans. They must have a sure-fire rain catchment system or know something I don't because a flat roof and rain don't mix."

I listened to Brad Pitt. He spoke with passion and vigor about the need for affordable and green housing for people living in the Ninth Ward of New Orleans, about areas hit hardest by the hurricane. But the grounding was flawed. Where was the input from people with the lived experience?

He and Clinton held an in-depth conversation about the project, how economically feasible it would make energy costs, the designs of the homes and that this was the right thing to do. There was and remains no doubt in my mind that Brad Pitt had the best intentions to show value to both low-income communities and affordable green housing with efficient renewable energy. But as the old folks say, "The road to hell is paved with good intentions." The project would eventually become the perfect petri dish for both green gentrification and blue lining.

The houses were built quickly to get folks back into the community and restore both stability and a sense of culture to an area ravaged by the hurricane. The housing designs were prepared by the best architects—in Japan, Germany and Africa. Upon first glance, they appear as a gorgeous backdrop to highlight the diversity of New Orleans. A mixture of shapes, textures and angles, the designs were beautiful. But they did little to take into account the climate and environment of New Orleans.

It's hot. And it's humid. And it rains a lot. And there is a lot of wind. And wind pushes pollution. And there are critters that get under and into houses. And on and on. By 2019, residents complained of unstable structures, faulty materials and mold. Residents were suffering from health issues that stemmed from the housing. Questions were asked about the expertise of those who designed and prepared for such an endeavor. Where was the community input? Pretty much anyone in South Louisiana, let alone the Ninth Ward, could have told them that flat roofs are a bad idea.

By 2020, several lawsuits had been filed and one of the houses demolished by the city. By the end of 2021, the Make It Right Foundation was entrenched in lawsuits on both sides, being sued by former residents while suing the people that were charged with making the whole thing happen in the first place. But the most frustrating part is what is happening to the people living there now. Those who cannot afford to move or lose the investment they placed into a new home are stuck. The idea of a "new" New Orleans is realized as old homeowners are bought out and replaced by investors who can afford to retool and rebrand the home as an "environmentally friendly" Airbnb. I feel bad because I really, really like Brad Pitt.

While the story of New Orleans's Ninth Ward and North Birmingham's plight may be similar to many Black and brown neighborhoods, there are solutions that are working across the country to resolve past environmental injustice and remove the barrier of access to homeownership, while considering the future of the planet all at the same time.

COMMUNITY LAND TRUST (CLT)

The Community Land Trust (CLT) model is becoming more integrated into affordable housing and environmental plans throughout America. The model is simple. An organization, usually nonprofit, purchases and holds a block of land in a trust. The CLT makes sure important community elements such as sidewalks, parks, sewers and gardens are incorporated into the land-use planning of the area. An interested buyer can purchase a house that sits on that land. The cost is affordable because the homeowner is purchasing the house but leasing the land, often for periods of up to ninety-nine years with an automatic option to renew. The CLT ensures that the cost remains affordable. The CLT homeowner is afforded the same rights and privileges as an ordinary homeowner. They can sell or lease and make improvements to the property.

There are currently over two hundred Community Land Trust zones in America and the number is growing rapidly. Northeast Farmers of Color Land Trust, California Community Land Trust Network and Florida Community Land Trust Institute are just a few examples of regional CLT programs across the United States.

CLTs are a good way to repurpose land previously considered environmentally uninhabitable. Former industrial sites, also known as brownfields, abandoned commercial areas and places where the population has diminished due to failed economic growth, are all candidates for CLT opportunities. Brownfields are eligible for federal funding to clean up and redevelop. Through a competitive process, a site can be considered for cleanup based on leveraged interest to create future economic and community benefits. An integration of affordable housing, commercial business development, small business growth and green space creates a multifaceted approach to both cleaning up past environmental hazards while also birthing new prospects for community expansion.

Make no mistake about it, these cleanups are costly and often do not hold the polluting entity accountable. Public and private partnerships, progressive financing and community innovation make the investment well worth the potential outcomes because they are more sustainable for the long term. If the community is vested and the business is partnered, it's more likely to last and adapt to climate changes in the future.

Dudley Neighbors Inc., located in Boston, Massachusetts, is an example of a neighborhood implementing this model. In the early 80s, the North Dorchester neighborhood of Dudley was suffering. Home to industrial parks and railroads, the environmental hazards associated with historically working-class and minority communities were alive and well. More than 20 percent of the neighborhood was vacant and the community's population of mostly young, minority people were surrounded by the elements associated with abandoned property—rancid smells, vandalism and a constant presence of trash. It was an eyesore for the city and concerned citizens decided to do something about it. An attempt at organizing the community to revitalize the neighborhood took place but failed to include those who lived in the neighborhood. Well-meaning white people were trying to fix a problem without the input of the people of color who were impacted. It did not go over well.

People did not give up. In the late 1980s, an organization was formed called Dudley Neighbors Incorporated, with a board comprised of the four main demographic groups of the community—Black American, Latino, Cape Verdean and white. The organization focused on equal power and representation in the CLT model. The organization collectively owned the property and was therefore able to maintain affordability and revitalization without displacement of people who had been disproportionately impacted by the downfall of the community in the first place.

Today, after thirty years, Dudley Neighbors is the site of almost one hundred affordable homes on environmentally clean property. They have created green initiatives such as urban farms, parks and green spaces as well as local small businesses that promote sustainability and resiliency for the neighborhood.

If the Dudley Neighborhood of Dorchester, Massachusetts, can figure out how to turn pollution into promise, others can follow suit. Models like CLT exist to create safe affordable housing without sacrificing people, land or the planet. These initiatives only work if we make the effort, and are creative and inclusive.

There is a saying—"Environmental groups move at the speed of light while communities move at the speed of trust."

Maybe then we can respond to Mr. Smith of North Birmingham by saying, "What are we going to do differently than others? We're staying right here with you. And we're going to repair and restore together."

MAKE IT MAKE SENSE—GREEN GENTRIFICATION AND BLUE LINING

Brownfields

Potentially dirty or toxic land. Brownfield is the term given to former industrial sites that have held some type of containment. Examples of common brownfield sites are dry cleaners and gas stations. If the site is to be repurposed for another use, then the ground must be cleaned or there is a possibility that those contaminants will impact the new owners without their knowledge. Cleanup is costly but the EPA (Environmental Protection Agency) provides technical assistance grants to communities looking to revitalize impacted areas and reduce the cost. Brownfields can become a huge economic boost to a community if the cleanup costs are covered.

Let's say there is an old tire plant in the neighborhood, and instead of leaving it as a vacant eyesore, the city wants to turn it into single-family homes, small businesses including a park and a grocery store. The venture would improve land use, tax revenue to the city and create jobs while also raising property values for the surrounding neighborhood. Nevertheless, the property must be cleaned, and this can cost millions that the city does not have. With the assistance of brownfield grants, the city can assist with cleanup and attract developers and businesses to invest in the project. Cities must share the cleanup cost, however, municipalities with a population under fifty thousand may be eligible for a waiver. In 2020, the EPA allocated at least $250 million for the brownfields program alone.

Community Land Trust (CLT)

Nonprofit organizations whose purpose is to own and redistribute land based on participation and advisement of the local community residents and businesses in the specifically zoned area. The idea is to build long-term wealth and community stability through homeownership and equitable distribution of land use.

Urban Renewal

Urban renewal was the plan to revitalize parts of large cities by tearing down old dilapidated buildings and areas considered slums, replacing them with more attractive, modern businesses and housing for a higher income group. This process included displacing many African American neighborhoods as well as large groups of diverse, low-income residents.

BEFORE THE STREETLIGHTS COME ON

1. Read the National Equitable and Just National Climate Platform. Become a member and find a local organization that makes equitable climate resiliency and adaptation a matter of practice and principle. Stay up to date.
2. Join or, at a minimum, attend and stay up to date on local boards and commissions that have oversight, influence or provide resources to community planning and development. Be the climate and environment person on the board. Opportunities for participation may include:
    a. City planning and zoning commission
    b. Insurance commissions

c. Community Land Trust boards
d. Realty/Realtor boards
e. Chamber of Commerce
f. City and/or County Redevelopment Commissions

3. Share the story of your neighborhood, family or community presence in the area and how it has changed due to climate and environmental impacts. Get creative and engage the young people of your family and neighborhood. Empower them to share their stories.
4. Identify a school and get involved with the long-term resiliency planning for future development in the community. It can be public, private, K–12 or secondary education.
5. Vote for people and policies that include local community members and culture as a matter of practice.

CHAPTER 7

THE CULTURAL APPROPRIATION OF COLLARD GREENS—FOOD INSECURITY AND THE CLIMATE CRISIS

"I do not like green eggs and ham. I do not like them, Sam-I-am!"

—The big yellow guy, *Green Eggs and Ham*

While visiting Sarasota, Florida, my husband agreed to join me in doing something I absolutely LOVE and completely unrelated to work—foot massages. We found a quaint little spot in a quiet retirement community. It wasn't overly glamourous—average size homes with small, neat yards, and the random golf cart illegally riding along the city street. With a few minutes to spare, we saw a shopping plaza with a health food-based grocery store and decided to stop in. It took less than two minutes for us to realize we were NOT in a regular store.

"Babe! You see this?" I tried to quietly whisper my shock as we walked through the fresh vegetable section. He smirked knowing full well where I was about to go. "What'd you expect? Look at where we are." I pretended to clutch invisible pearls. "Dex, this bundle of collard greens is $5.99! Five whole dollars and ninety-nine whole cents!" I held up the thin bundle of organic greens by the stalk. There was more stalk than green, but the wrapper on the green twist-tie proudly proclaimed, ORGANIC AND LOCALLY GROWN. "Dude, there are only three leaves on this thing. What in God's green earth can you do with three collard green leaves? Who is eating THREE greens?!" Without missing a beat, my spouse, best friend and partner in crime pointed to the tofu in the next section of the store and responded, "Somebody trying to make a collard green wrap with this stuff!" I burst into laughter.

Both my husband and I were born in Greenville, Mississippi, a place frequently referred to as "the heart of the Mississippi Delta." We playfully banter with each other about who spent the most time in the gardens of grandmothers and godmothers. He always wins—his grandfather had a farm with pigs, chickens, a tractor and rows and rows of homegrown vegetables. He gets PTSD whenever anyone speaks of picking, let alone eating, okra. Whether they be mustards, collards, or turnips; greens have always been a staple in our diet. Sunday dinners were not complete without a pot of greens. Coming home from college with friends? They got greens in a carefully frozen, lovingly foil-wrapped dish to take back to school. At every funeral repass, there was a section of the white paper plate reserved for greens, cornbread and if you're lucky a sliver of hamhock. Dexter and I picked greens out of backyards, bought them off the back of a flatbed truck pulled over on the side of the road and have stripped leafy greens from their stems. We can eyeball how many bundles of greens it takes to feed a family of four or a family reunion of one hundred. Between the two of us, we have well over eighty years of "greens" experience and like two middle school kids, we spent the rest of the time in the store trying our best not to draw attention from the store clerks as we snickered, giggled and snorted about the insanity of a $5.99 bundle of collard greens.

My husband's assessment of the "who" and "why" of $5.99 collard greens was not far from the truth. The rhetoric often heard from environmentalists and climate advocates is that moving to a vegan or vegetarian diet is one of the best ways we can take individual action to save the planet. We've already established that a major factor of global warming and climate change is human activity that increases heat-trapping gases. If we're going to be effective in fighting climate change, then we must reduce this activity in every way. After fossil fuels, deforestation and food production are the biggest emitters of heat-trapping gases. We cannot achieve climate stabilization without changing entire food systems from beginning to end—feeding livestock complete through package and distribution channels. Individual action alone won't cut it. To avoid further catastrophic impacts from global warming like wildfires, floods, heatwaves and drought, we have to work together to change how we produce food while cutting our meat intake at the same time.

CLIMATE IMPACTS TO FOOD PRODUCTION

Unfortunately, the process isn't that simple due to impacts from climate change that are already underway and directly affecting the existing food security for vulnerable populations. As a result of climate changes and weather shifts, the increased floods and drought we experience today negatively affect the ability of farmers around the world to grow and produce food. Absent of immediate action, weather events will worsen and disrupt crop growth. Global food production will decrease and food costs will rise.

A perfect example took place in Mississippi. From 2019 to 2020, the Mississippi Delta region experienced one of the longest flood periods in history. With the river swollen, flood waters grew and stayed—over crops, fields and roadways—until there was more water than the land could handle. The flood created an upstream domino effect. Barges carrying agricultural goods from farmers in Iowa could not safely navigate the swollen river flood waters and were in turn, unable to ship down the river.

Turned away from the river barges, farmers turned to trains which were quickly overrun and weather impacted. When transportation routes halted, product was stored in silos. When the silos became full, corn rotted in the fields. At the same time, the price of food products related to this system, like animal feed, increased. The production expenses were passed along the supply chain so the cost of food products, including chicken and beef, began to rise.

As consumers, we wondered why the ground beef was a dollar more than it was last week. But in reality, the entire food system had experienced a financial system shock as a result of a flood. When extreme weather events increase, the cost of producing food increases and the cost we pay in the grocery store increases. No one feels this burden more than low-income households that have no idea of the climate change affiliation to rising grocery store prices, let alone how to accommodate for the high cost of food due to climate change. The process exasperates what we have come to know as food deserts.

THE ENVIRONMENTAL POVERTY TAX OF FOOD DESERTS

Being poor is expensive.

People who live in food deserts know this fact well, wherein urban areas a supermarket or grocery store is more than a mile away from the central population and more than ten miles away from the central population in a rural area. It's called a desert because living food items like fruits and vegetables are hard to come by in these regions. It's the corner store in the Black neighborhood where you can buy chips, a soda and cigarettes, but no apples, tomatoes or greens. It's having a Walgreens or CVS nearby where you can get eggs, milk and sugar, but not carrots, cauliflower or cabbage. It's living in a community in which you either need a car or be willing to take several modes of public transportation to purchase semi-fresh fish and meat from a deli.

People living in food deserts carry the burden of paying more for their food. The additional cost of time and transportation to get the food must be calculated and then added to the final food bill. A family or individual able to walk to a neighborhood grocery store and purchase a bundle of greens for $1.99 pays less than the individual who spends money on gas or public transportation fare, plus the extra travel time to get the same bundle of $1.99 greens. This amount doesn't account for added stressors like childcare, education and public safety. Do you know how much time is saved if you can grocery shop while the kids are at school versus having a three-year-old in tow? I do. I can run in and out of my local Kroger for milk and bread in seven minutes, thirty seconds flat and that's IF there is a line at the self-checkout. If I have kids with me, that same store run takes three hours and twenty-two minutes. Think about the added stress if you cannot read the signs and labels because they are not in your native language or you never learned to read. How would you feel if every time you walked into the grocery store an armed officer was standing guard outside because of safety concerns? According to the FBI, seventy-eight people were killed in incidents of grocery store mass shootings between 2000 and 2020. In 2021 alone, we saw the deaths of fifteen people, including one child, in grocery store shootings in Florida, Colorado and Tennessee. In 2022, our country was rocked yet again by the horrendous mass shooting of eleven Black people and two white people in a Tops grocery store in Buffalo, New York. With the increase of gun violence and random shootings in places as common as grocery stores, nutritional value and making good choices be damned—folks just want to get in and out ALIVE. Some may disagree but part of the value of locally owned and operated grocery stores is knowing and being part of the neighborhood. The corner store might be sketchy but at least I know the guys on the corner. They know that I know that they know, if anything goes down they've got my back. But even that comfort was jilted after an eighteen-year-old white man, who claimed he had to execute this attack for the sake of the white race, drove over two hundred miles just to shoot Black people in their own neighborhood grocery. Each of these elements adds to the difficult process of accessing healthy food and cannot be excluded from "why" chronic health problems like diabetes and high blood pressure develop at a rapid pace within low-income and often Black communities. All of this is made worse by the rapid impacts of climate change—less food production equals higher costs, less access and additional social stressors to obtaining what food is available.

That brings us to the question, "Well, what *is* available to eat?" Under the scenario above, food that is available and low cost, is often low quality and not the best for anyone's overall health let alone those who suffer from chronic health problems or are immunocompromised. Food oppression is real—it's not that residents in low-income communities or Black neighborhoods don't want to eat better, it's about what foods are within reach. When it's easier, cheaper and safer to walk to McDonald's and feed a family of four from the dollar menu than cooking, you're nutritionally oppressed. Food deserts often overlap as environmental frontline and fenceline communities. They are the same neighborhoods and communities battling environmental injustices such as not being able to drink the water out of the tap because it has lead in it, or the inability to play and exercise freely outside because the air pollution is so bad it causes an asthma attack. Add the additional cost of food accessibility, affordability and nutritional value and the dollars add up. Without a doubt, residents of food desert areas may be below the poverty line but pay more for food. This line of separation will only widen as climate change worsens.

The Supplemental Nutrition Assistance Program (SNAP) is meant to remove a barrier to access nutritional food by providing additional money to low-income families. Most people know it as the Food Stamp program, and it's been around since 1939. Today, an "EBT" card (Electronic Benefits Transfer formerly known as food stamps) is the means to purchase food on the program, but holders of an EBT are the same people paying what I call "EPT"—the "Environmental Poverty Tax." This is the cost assessed to those living in an environmentally toxic community, fenceline or frontline, with no means or assistance to correct, improve or leave. The environmental poverty tax shows up when people who need an EBT to supplement their food budget are the same people without access to clean, drinkable water so they have to buy bottled water for consumption or to add to baby formula. The cost of access to clean drinking water increases when poor people are charged extra to not only obtain drinkable water, but also pay for the additional costs of plastic bottles while bearing the brunt of living in polluted communities stemming from the petrochemical industry that creates it. EPT (Environmental Poverty Tax) appears again when residents must purchase special sensitive skin soap (on top of the water they need to bathe in) because air pollution presents irritants that cause or make conditions like eczema worse. The EPT is evident when folks have to buy pots and soil or raised beds to plant greens because the ground itself is too toxic to eat from. Without the inclusion and recognition that poor people have and will continue to bear the brunt of climate impacts to our food supply and how we consume it, mass transformative climate solutions will not be embraced and implemented by society. We must collectively think of how we can remove the environmental justice barriers that prevent the establishment of equitable access to the basic necessities of food including clean water, air and the land to grow it.

SPEAKING THE LANGUAGE OF FOOD DISPARITY AND CLIMATE

Approaching the climate crisis and food insecurity is a sensitive yet complex matter, best discussed through trusted sources with lived experience in the Black community. Somewhere between the history of enslaved Africans, Indigenous people and agriculture in America lies the intersection of protecting both the planet and the people on it. Despite systemic racism, the ancestors were supernatural, armed with the skills to transform bits of nothing into a way of life. It begins with abandoning the stereotypes and dispelling the myths. Black folks have always eaten our greens and vegetables. From singing sensation Beyoncé to the NFL's David Carter, celebrities, athletes and musicians in the Black community talk openly and joyfully about living a plant-based lifestyle. In 2020, vegan enthusiast and TikTok sensation Tabitha Brown revised her role as a struggling actress to an overwhelmingly successful social media influencer with bacon-flavored carrots. I will admit, I was skeptical at first but there was something about the soothing way she sprinkled garlic powder on carrots with such love that it sounded like an incantation of the ancestors saying, "Thank you for remembering the old ways, we are pleased." By 2022, Tabitha Brown's ethnic vegan lifestyle was not only embraced but profitable in everything from books to a clothing line at Target.

In 2016, the Pew Research Center found that Black Americans are more likely to identify as vegetarian or vegan compared to all other Americans. In fact, a third of Black Americans are cutting back on their meat intake versus one-fifth of white Americans. While the United States remains as one of the top meat-consuming nations in the world, among the Black American populace who are reducing their meat intake, the reasons most listed are to "improve health" and "the environment." Marketing teams are taking note. Burger King was the first to debut the Impossible Burger. Kentucky Fried Chicken has a plant-based chicken option. Even with the knowledge of these polls and facts, mainstream veganism, similar to mainstream environmentalism, is largely considered as being founded, maintained and grown by white people.

The mainstream movements borrow many ideas from people of color. Black experts with lived experience can be a trusted voice to other Black people when it comes to a food lifestyle that is not only germane to Black history, culture and future existence but is central to understanding the climate crisis. So often, these voices are excluded and disenfranchised from participating in the conversation. In 2014, the vegan food site, Thug Kitchen, faced well-founded accusations of cultural overstep. At face value, it appeared to be a Black American vegan space, full of Black American vernacular, ideas and community. In reality, it was run by a white couple that used Black American terminology to gain an audience. Authentic Black vegan food experts described the debacle as cultural food appropriation. As one expert put it, "People rarely go to the second page of the Google search."

This exclusion leads to further flaws in how the message of food and climate is delivered. For Black Americans, there is a historic intersectionality of climate and environmental issues with equity and social justice issues that can and should be addressed by tackling food disparity. The two cannot be untangled. Come at us with the wrong tone or inability to embrace the intersection and the entire message gets thrown off track. Compassion and action for the planet cannot exist without compassion and action for the people on it, including the underserved and marginalized. How and why is a person of color expected to resonate with white, mainstream, environmentally moved, vegan supporters of "cruelty-free" eating for the sake of the planet in the future, when at times, these same people do not exhibit the same sense of compassion for the hunger, suffering and cruelty of Black people by the police? They are not two separate issues but instead two ends of a knot tangled by a history of systemic oppression and racism that overlaps food security and climate.

It is difficult to process the words, "You should eat more vegetables because it's good for the planet" as it's often misguided and ill-received when it comes to Black communities. It ignores our existing cultural affiliation with plant-based living. It does not take into account the historic love-hate relationship between African Americans and agriculture through the American period of enslavement, nor does it account for post–Civil War systemic barriers that prevented us from doing the very things we were told are good for the environment and our bodies. It almost sounds like a chastisement for not having sense enough to do something so obviously good for humanity when in fact, we've been well aware of the benefits but have been blocked from doing the work for ourselves first. How can I plant a garden in the backyard when years of systemic racist housing policies have prevented me from owning the property where the yard sits? IF I am able to plant a garden, how do I keep it watered when the water source isn't fit to drink? Most stinging is the self-inflicted victimization it presents to us, assuming that the economic and environmental position with which we find ourselves is of our own doing. The people paying an environmental poverty tax for bottled water because access to clean water doesn't exist for them aren't too happy when told by mainstream society that they should recycle, not use so much plastic and be responsible for where those bottles end up. The mood becomes even worse when they hear the often-repeated statements that not only associate reducing meat intake with automatically reducing chronic disease and other disparities that disproportionately impact Black people, but also climate change and therefore, something that the Black community should want to gladly do. It's someone saying, "But if she would just stop eating bacon, your big mama can lower her blood pressure *and* do something good for the planet at the same time. That's good, right? We all want to be healthy and do what's best for her and Mother Earth, right?" What mainstream environmentalists fail to realize is that while the best intent may be implied, the message we hear and respond to reminds of me of the scene in *The Color Purple* when the mayor's wife, Miss Millie, asked Sophia if she wanted to come and be her maid.

"Hell Naw!"

"What did you say?"

"I said . . . HELL NAW."

"We" did not all poison the planet equally so why are "we" being told that "we" must change our lifestyle to accommodate something we had little part in screwing up? "We" did not place our homes next to industrial waste sites and on toxic land by choice. The reason we're eating what we eat is born from the traumas of slavery and the plain "old school" magic our ancestors accomplished with what was given to them and the cultural accommodations made to adjust to what we had. "We" didn't always eat this way and the path back to wholeness certainly will not be led by those who caused the problem in the first place. We are the big yellow guy in Dr. Suess's *Green Eggs and Ham*, being followed around by Sam-I-am and constantly asked if we want the GREEN eggs and vegan ham in the hopes that constant pressure will yield a willing but guilt-filled participation. The whole conversation is tantamount to an abuser telling the domestic violence victim, "I know I beat you really bad, but if you help me pay for the counseling and agree that we won't put our hands on each other, I promise life will be better." As my mother would say, "The devil IS a lie, we're not falling for that one."

The good news is that there is a middle ground where justice and anti-racism working together has the potential for tremendous and expedient benefits if everyone is willing to listen. That middle ground is called climate justice and it is spreading.

TRANSFORMING CLIMATE MESSAGING INTO MEALS

Between the history of enslaved Africans and agriculture in America, and the intersectionality of lived experience overcoming poverty with plant-based lifestyles, Black Americans are primed to provide wisdom, insight and innovation to feed our country's people. Remember the "old school" skill of transforming little of nothing into something magically delicious? It is derived from generations of turning scraps into seasonings, rations into refreshments and sprinkling mustard seed faith to feed multitudes. Think about it—most historically Black colleges and universities were founded as agricultural and mechanical training institutes designed to educate and transition the Black labor class into the industrial era. How many of us played the role of George Washington Carver and his peanut farm at Tuskegee as part of the elementary school Black History Month program? Evidence of African American connections to the environment and land through food production became stronger with the expansion of land-grant colleges for Black students. The 1890 land-grant institutions program established state-affiliated Black colleges and universities that focused on research in the areas of agriculture, forestry and food. Under these programs, historically Black colleges and universities develop innovations around food production, renewable energy, biotechnology and food safety to this day.

Luckily, the shift is happening as all environmental leaders, regardless of race, class or creed, recognize the short time left to take bold implementable actions on climate that can innovate and balance our food systems before time runs out.

SEAWEED MIGHT SAVE THE HAMBURGERS

Enter the seaweed diet. But for cows.

I was sitting at TED Climate Countdown 2021 in Edinburgh, Scotland alongside Dr. Beverly Wright from the Deep South Center for Environmental Justice and Mrs. Peggy Shepard, co-founder and director of WE ACT for Environmental Justice. Both women are tremendous figures in the environmental justice movement and for years I admired them from afar but it wasn't until later in my career that I got to know them better and love them all the more. Peggy Shepard is the epitome of grace, style and class encased in a petite but solid body—and carrying a smile for the gods. That same smile can melt iron in seconds when crossed by someone dismissive of her expertise and place as an environmental justice expert. Peggy can say, "Go to hell" by nodding her head and you'd automatically know to ask Satan for your ticket. On the other hand, Dr. Beverly Wright will actually tell you, "Go to hell and pick some flowers along the way" out loud. In a room full of people. Repeatedly. Then she'll write about it, publish it and call you to read it to you again just in case you missed it the first time. With over forty years of environmental justice expertise, Dr. Wright is tall, bold and outspoken. Together the lived experiences of these two Black women fighting and advocating for environmental justice would fill volumes of written work. They are fierce and feisty and I'm blessed to call them mentors.

During this particular TED climate session, we'd heard about how the world "must" shift to a plant-based diet if we are to survive the impending climate crisis and the eager applause from a heavily vegetarian audience was beginning to get old. This, accompanied with the fact that every meal at the event was plant-based, had worn its welcome on those of us who liked a little smoked bacon every now and then. "I'm not giving up my hamburger, I don't care what the hell they say," proclaimed Dr. Wright.

"Ummm hmmm," said Mrs. Shepard.

Even in the dark auditorium, their faces gave that "I wish you would" stare that comes from an auntie that you've made one too many smart-ass remarks to and she needs to remind you that you are far from grown and taking it too far. I moved not a single muscle.

The next person to enter the stage was Dr. Ermias Kebreab, Associate Dean of Global Engagement at UC Berkley and Director of the World Food Center. For the next eleven minutes, this astute Black man from Ethiopia with an amazing presence of peace and a smile walked us through an innovative process of changing livestock feed to a seaweed-based diet. I sat forward a bit to make sure I was hearing correctly while Dr. Wright shifted and said, "Well, finally somebody with some sense, now we're talking!"

The biggest problem with meat from cows is that their natural process for digestion produces a lot of methane gas, and I do mean a lot. According to the EPA, over a quarter of the methane gas in the world comes from cows burping and passing gas. Since methane is one of the biggest contributors to climate change, if we cut what's causing the gas we slow global warming. One way is to reduce the number of cows in existence, that is, cut back on meat. While that is one solution, Dr. Kebreab presented another idea—why not just reduce what causes the methane in the cows in the first place? Livestock scientists found that a particular red seaweed, common around the world, contained certain compounds that prohibited the stomach from creating methane. When incorporated into the feedstock for cows, scientists noticed a remarkable decrease in the amount of methane produced by the animals. The scientists basically created extra strength Gas-X for cows. Dr. Kebreab excitedly shared how in some studies the methane production in cows was reduced by over 90 percent and there was no difference in the taste of meat. I sat up even straighter. This type of transformative shift in livestock feed could mean keeping hamburgers, but also reducing the cost of food production while inserting another new job industry for working people. But there is much work to be done and here is where we enter. What better opportunity for HBCUs to partner and help scale while normalizing and being innovative around the seaweed diet for cows? There are problems of process to be solved: collection of seaweed, scale of process, location, distribution, etc. But these are all areas Black people have significant history and success (Remember George Washington Carver figured out over three hundred uses for a peanut; I can't wait to see what else we can come up with for seaweed). As I sat with Dr. Wright and Mrs. Shepard, now with pleased looks on their faces, I realized Dr. Kebreab gave new meaning to the words "Surf & Turf" and it just might be part of the climate food solution to save us all.

The food-related climate challenges we face require creativity that reforms farming and agricultural processes, utilizes renewable resources to increase crop growth while removing carbon from the atmosphere and identifies ways to reduce our meat intake while providing nutrients to the body. Each one of these challenges has already been faced and overcome by our ancestors during slavery, reconstruction and the Great Migration to the north. But it will take partnership, allyship and inclusion from the mainstream environmental movement. What's exciting is that it is already happening. It's creative, full of energy and innovation. The great part is that it won't cost $5.99 per bundle of collard greens to make it a reality.

MAKE IT MAKE SENSE—CLIMATE AND FOOD SECURITY

Food oppression

The effect of a food-related policy that disproportionately impacts poor people such that they pay more for food.

"EPT"

The "Environmental Poverty Tax." This is the cost assessed to those living in an environmentally toxic community, fenceline or frontline, with no means or assistance to correct, improve or leave.

BEFORE THE STREETLIGHTS COME ON

1. Plant, grow and farm. Utilize regenerative and restorative farming practices that are beneficial for people and the planet. Foot Print Farms and Soul Fire Farm are two examples of organizations committed to food and farming equity. Leah Pinneman's book, *Farming While Black,* is a must-read manual for everything from reclamation land practices to restorative agricultural policy and financing.
2. Look for ways your community may be unknowingly paying the EPT—Environmental

Poverty Tax—on food and water. Work to remove barriers that increase the cost of food for environmental justice communities. Invest in and advocate for clean water sources instead of constantly donating water to communities with unhealthy water supplies.

3. Find creative alternatives to meat and meat byproducts. Spice it up. I didn't say cut it out altogether, but look for ways to cut back on the bacon. Patronize Black plant-based businesses, influencers and creators. Create and share plant-based recipes. We are a creative people!

4. Support HBCUs, specifically agricultural and mechanical schools with a focus on food production and equity. Florida A&M, Tuskegee Institute and Alcorn State University are great starts. Encourage students to enroll and participate in career paths that include all aspects of the food supply chain. Donate to HBCUs.

5. VOTE for officials and policies that support equity in food production and distribution.

CHAPTER 8

TOO HOT TO LEARN, TOO COLD TO CARE—THE EDUCATIONAL IMPACTS OF CLIMATE CHANGE

One thing that has never changed about starting school in Mississippi is the heat. As with most southern states, school traditionally begins in August and releases in early May. These time frames were in place to accommodate the farming season by allowing kids to return home and help out in time for planting. For me, the school day started as early as 7:15 a.m. and we attended what was called "60 percent days." We were out at or slightly after noon, avoiding the hottest part of day. The reason was simple: in the mid-1980s, most schools in the Mississippi Delta did not have air conditioning.

It must have taken a special kind of teacher to educate in that environment. Now that I have children of my own, I understand that they were not merely teachers, they were saints. I laugh now and wonder how in the world those teachers lived with the energy, sweat and smell of a classroom of kids. I used to hear, "Y'all smell like outdoors" so often that I thought it was a fragrance. Truth be told, I know we must have been something awful. But if you were a Black student attending public school in the Delta, chances were your school had received no major infrastructure or building upgrades and enrolled few white students, if any at all. Most schools in the Mississippi Delta were built during the early 1900s, updated in the 1940s and 50s, then later abandoned by white leaders upon integration and the onset of private schools and academies built for white students. There has been no new public school construction in the Mississippi Delta since the early 1980s.

The old public school buildings were structured to best promote air flow since central air conditioning was not a standard of building construction. Wide breezeways connected each building with large plots of green space in between. The classrooms were situated in blocks with an exterior door to each room and the shared hallway. The halls were tiled with cool shades of blue and green in alternating patterns while water fountains were located in every hall and they usually worked. Teachers could open both sets of doors and get a pretty nice although slightly warm breeze to flow through the classroom. The hum of window fans was as common as the sound of cicadas in the summertime. By the time I entered fifth grade in 1985, some classrooms were updated with window units. The idea of teaching and learning in the stifling heat of a late Mississippi summer meant having patience and understanding that grades would get better as the days got cooler. It wasn't until I was in the sixth grade that the thought of going to an air-conditioned school became an idea close to reality. But at age twelve, I had no idea it was because of where I lived.

SCHOOL ZONING AS AN ENVIRONMENTAL BARRIER TO EDUCATION

Zoning has historically and systemically been used as a method to separate communities based on race and class and drive investment into wealthier white educational systems. It coincides with a long list of barriers—redlining, environmental waste permits and highway transportation routes—as an effective means of discouragement, limitation and divestment in Black school districts. As integration made its way across school districts in America, it was accompanied by white flight, thus leaving once well-maintained, well-financed schools at the mercy of a diminishing tax base. In Mississippi, private academics and schools were created en masse across the Mississippi Delta. Mississippi is rural and in the 1960s and 70s, there were no suburbs in urban areas. To physically move would mean giving away three things: land, money and political power. Instead, white parents and leaders used these elements to create new private schools not bound by the zoning regulations and requirements of public entities. Making matters worse, white state leaders changed the laws in the state of Mississippi that allowed for the passage of local bonds to build new public schools. Instead of needing a simple majority vote, new laws required a local school bond issue to pass by 60 percent. As the white population of students within the public school system declined, the maintenance of the schools declined in tandem. Public school boards, manned by white school board members, held the authority and ability to withhold funds. In turn, Black communities were subject to not only subpar streets and sewers that fed school districts, but also schools locked in the pathway of poor air quality, little maintenance and at the mercy of climate change.

My school system was no different. By the time I entered the public school system in Greenville, our town of less than forty-eight thousand residents housed five high schools—four of which were private or parochial.

In the late 1980s, Greenville, Mississippi, had two public junior high schools and students attended the school zoned to their home address. Solomon Jr. High School was the "white" public school and the most recently built in the city. Built in 1967, it was a standout structure. While most of the other schools were built in web design—separate buildings connected by breezeways and corridors to the main building—Solomon was one towering structure with no windows and very few exterior facing doors. On the opposite side of town sat Coleman Jr. High School. "Coleman," as it was referred to, was the former Black high school in Greenville named after beloved educator Lizzie Coleman. In its heyday, Coleman High School was the premier school for African Americans and produced some of the state's top Black doctors and teachers. When schools were desegregated in 1969, Greenville, Mississippi, was one of the first districts to openly embrace integration and had prepared a plan to enact it. Coleman became a junior high school, conveniently zoned for most of the city's Black youth while Solomon was the junior high school zoned for most of the city's white communities. Solomon had air conditioning. Coleman did not.

In 1988, Solomon Jr. High School housed the Greenville public school system's gifted and talented program for middle school students. When I attended during the late 80s, Solomon was 60 percent Black and 40 percent white, a stark difference from our cross-town rival Coleman, which boasted a student population that was well over 90 percent African American. Both Solomon and Coleman had "affirmative action" rules. On the homecoming court, there had to be one Black and one white maid for each grade. If the student body president was white then the vice president had to be Black and vice versa. Solomon boasted better academics and better test scores. Teachers with higher degrees and more experience were thought to teach at Solomon. Coleman had the better football team and band. In my sixth-grade mind, the smart kids and the white kids went to Solomon in the air conditioning. Everyone else went to Coleman.

EXTREME HEAT AND EDUCATION

Students do not learn well in the heat. A study in the *Nature Human Behaviour* journal examined the widening educational achievement gap between Black, brown and white students and its relationship to extreme heat. The study is part of a growing body of research and evidence that proves not only do test scores drop, but it's also disproportionate in Black communities and it's related to temperature conditions both in school and at home. Black and brown students that attended schools subject to warmer temperatures due to their geographic location, experienced lower test scores than their white counterparts. While the study was conducted across fifty-eight countries, the researchers found a unique difference in the United States. The impact of heat was only seen in Black and Hispanic students. For every day of eighty-degree temperature or greater, students fared worse on standardized tests.

The study concluded that the main culprit was rising temperatures but also failing infrastructure in Black and brown communities. Climate change creates extreme weather conditions that exacerbate heat, but communities comprised of primarily Black and brown students lag behind when it comes to investing and updating school infrastructure. The lack of investment directly impacts the learning environment for students.

This is especially true for schools that lie on the frontlines of climate change. Schools in New Orleans have to account for the additional cost of maintaining underground infrastructures such as water lines and pipes that are affected by relentless floods and ground saturation. Schools in places like Port Arthur, Texas, can barely rebuild from one hurricane before the next one is spotted in the Gulf. In 2017 Hurricane Harvey destroyed two schools in Beaumont, Texas, but it was not until 2021 that they were able to break ground on the new buildings. These circumstances are in addition to the fact that minorities are more likely to live near toxic waste sites, exposing them to hazardous chemicals during floods and storms or industrial facilities exposing them to harmful emissions. Public housing developments full of concrete, asphalt and stone—all materials considered impervious—hold heat. Green spaces, trails and trees allow for ease of airflow and also refurbish the atmosphere. Access to basic funding for majority-Black schools or those located in Black American communities, belongs to a long list of racial inequities that magnify the effects of climate change.

A study similar to the *Nature Human Behaviour* journal considered the prevalence of mercury pollution within Black communities and found greater exposure and particular dangers to children and women of childbearing age. Acute or chronic mercury exposure is dangerous to both children and adults, but children are especially vulnerable. Mercury poisoning causes brain degeneration and growth impairment in fetuses. The researchers at Jackson State University, a historically Black university (HBCU) in Mississippi, identified Black children as markedly more vulnerable to mercury toxicity, and the results could include developmental impairment of the central nervous system causing educational delays and disability. One of the first things an expectant mother hears upon discovery of her pregnancy is to limit the intake of fish, especially grouper and tuna because they carry high levels of mercury. It collects in the fatty tissue and in a pregnant woman it can travel through the umbilical cord and the placenta, eventually reaching the baby's brain. But it is also important to understand how mercury gets into fish in the first place and why African Americans are more susceptible. Coal-fired power plants are the main and largest source of mercury through emission, accounting for over 40 percent of mercury pollution in the United States. When power plants release emissions into the air, elements like mercury are carried through the air and fall as part of the rain cycle into rivers and streams. For this reason, every state in the United States has some type of fish advisory to ensure people are aware of potential mercury contamination from eating the fish. Contaminated fish is eaten by other fish, animals and humans.

When taken into consideration alongside data outlining the close proximity of African Americans to coal-fired power plants, it's no wonder that Black people are more likely to be subject to mercury poisoning and as a result have higher rates of education disparities due to pollution. According to the NAACP, over a million African Americans live within one mile of a coal-fired power plant or an oil and gas operation. Looking around urban neighborhoods and the places these plants are located, particularly along the southeast and in coastal regions that have a higher rate of African American neighborhoods, it's clear: African Americans live not only under a cloud of pollution, but this same pollution impacts the ability of African American children to learn at the same rate as their white counterparts.

Mercury poisoning and extreme weather are two examples of environmental elements that impact education. Pick any region of the continental United States, and we can find links between environmental injustices and education. Being Black in America is the top criteria for susceptibility to lead poisoning—not poverty or housing, but just by being Black. Lead poisoning is dangerous for children as it damages the brain and harms development. Black children living below the poverty line are four times more likely to suffer from lead poisoning than their white and Latino counterparts. A look in the inner cities reveals housing developments that have not only been redlined to group a large number of people into small spaces, but are also subjected to ineffective permitting requirements that have not been enforced on emitting agencies that are in those same areas. Chicago, Detroit, Philadelphia, Pittsburgh and Columbus are examples of urban communities that are dealing with the real-time issues of environmental poisoning and lack of infrastructure for inner-city development in schools. If we look to the rural South, places like Louisiana, the Mississippi Gulf Coast and Alabama prepare themselves to be resilient from yet another storm while repairing their schools from the previous year's disasters. To make matters worse, southern coastal communities are fighting the growth and contamination of petrochemical facilities, one of the largest growth industries in the world. How are children who live with a constant threat of an increased climate crisis while neighbors to constant pollution supposed to continue their educational advances at the same rate as their white counterparts?

Similar factors extend to higher education. Historically Black colleges and universities (HBCU) and minority-serving institutions (MSI) are held in high esteem as centers of academic excellence for the Black community and trusted centers of information for crisis response. The majority are located in the southeast, within Black and brown neighborhoods, and serve students that come from a diverse demographic across the country. But HBCUs fall far behind their white counterparts in facilities and endowments. When storms hit schools like Prairie View A&M University in Texas or Xavier University in New Orleans, the school serves its students and community. It becomes a shelter, a food source, and an information hub. While expected to be everything to everyone in the community, the investment necessary to keep up with demand is non-existent. Somehow, they do it anyway.

We cannot believe our country can adequately prepare for climate change without a serious investment into the infrastructure of public schools and minority-serving institutions of higher learning. COVID-19 responses have shown us what is possible: innovative ways of learning that can be adapted to the environment of the children served. Through required distancing, we've learned that classrooms can be effective outside where students can experience the environment while learning. Research has shown that student test scores improve as well as their health and behavior. Making sure students living in highly polluted areas have the same access should be an environmental priority as well as an educational focus. Industrial permitting requirements and renewals around educational facilities should be the focus of the PTA, president and the city council. School buildings and renovations must have climate resiliency and adaptability in future planning. While there is no catch-all blanket that can be thrown over every educational problem, we must start by asking the community first: what do your families need to ensure the success of your students? We'll find that solutions to environmental injustices in the Black community go hand in hand with improving educational opportunities.

In 2021, the *New York Times* printed a story about the tragedy of rural schools and highlighted Holmes County School District in Mississippi, not far from where I attended school in the Mississippi Delta. The article's words spelled out my own experience over thirty years ago. "Students may grow distracted if their classrooms are too hot or too dim to make out the board, and schools with poor ventilation may leave children drowsy as hundreds of teenagers exhale carbon dioxide into the air. In the most dire situations—settings like the ones Holmes County students sat through every weekday—mold can sicken teachers and students enough to miss class." It pains me that today, we still have entire school districts without air conditioning, ill-maintained buildings subject to flooding and a failure to acknowledge the relationship between these basic infrastructure needs and educational success. But with the looming climate crisis, what we know and have experienced of hurricanes, floods and wildfires, and what climate scientists have forewarned us of what will come, we have no time to neglect the spaces of learning for the next generation of problem solvers. Equity is not an option and barriers to adequate structural funding for schools must be removed. It does not matter if those barriers are political, social, racist or religious; the clock is ticking on all of humanity. We may not have the answer to solving the climate crisis, but we damn sure shouldn't preclude our children from the opportunity to learn basic principles that could lead to the climate innovation that will save us all. If it means we collectively hold umbrellas over their heads as they study under leaking roofs—so be it until we can fix the roof. If it means we bring box fans to cool them as they learn the mechanics of electricity and grid stability—let us all line up at the local Dollar Tree or Walmart and bring the fans. More importantly, if it means that more of us need to run for school board, state legislature, county council, serve on boards or commissions that direct policy, then let's get to it. We can be more than band parents and football team chaperones. The education of our children is on the frontlines of the climate crisis and the only barrier of protection they have,

MAKE IT MAKE SENSE—EDUCATION AND CLIMATE TERMINOLOGY

Affirmative Action

Affirmative Action is a policy to create opportunities for people who would have no way of being included or provided the opportunity if not for a specifically targeted program. Members of minority groups, including but not limited to Black, Latino and Asian Americans, women and the disabled, are often the targets of affirmative action programs to create greater diversity and equity in various programs.

HBCU

Historically Black Colleges and Universities are institutions established before the Civil Rights Act of 1964 and have the education of Black Americans as their primary mission. While open to people of all races, HBCUs were created to address the lack of educational opportunities for Black Americans. The first HBCU was Cheyney University of Pennsylvania, founded in 1837. Today, there are over a hundred HBCUs boasting a collective enrollment of over 200,000 students.

Lead Poisoning

Lead is a natural metal found around the world but when it comes into contact with the human body, it's toxic. It's dangerous to children because kids absorb it easier than adults and are often in closer contact. Babies crawling on the floor, toddlers eating dirt, preschoolers chewing on pencils, are all ways that young people come into contact with lead. Children who live in low-income communities, often in houses built before 1978, are more likely to get lead poisoning. Lead poisoning can cause development and educational delays as well as significant harm to the body's organs.

Private School Academies (Segregation Academies)

These are schools that were established after the landmark Brown v. Board of Education Supreme Court case that declared segregated schools as unconstitutional. Many of the schools are located in the southern United States and were founded by white parents who did not want their children attending school with Black students. The prevalence of private schools and segregation academies in the South created de facto racial segregation in public schools. By default, public schools became majority Black American.

BEFORE THE STREETLIGHTS COME ON

1. Participate in your local school board and ask for the long-term planning for climate adaptation and building resiliency. Are classrooms well ventilated? Does your school have central air and heat? Is there freedom of airflow?

2. Look for energy efficiency opportunities in your school or educational institution. We can do more than change light bulbs to energy-efficient lighting. White roofs can help reduce the heat absorbed in a building and lower the cost. If new buildings are planned, are they taking advantage of new technology to lower both the cost and overuse of electricity?

3. Create a proof of concept plan for renewable energy power to be used in at least one building at the school. A student-run solar farm is not only educational, but also a lesson in business, shared resources and science.

4. Explore outdoor classrooms and opportunities to get students outside and into the elements of air and nature if possible. Advocate to remove barriers like air pollution or provide protection in extreme heat. Identify a portion of the land to grow food to be used in the school. Create a class and teach students farm-to-table practices. Design natural safe spaces that include air, water and land.

5. Plant trees indigenous to the area along the sidewalks and driveways to the school building. Trees are natural filters and over time will reduce the amount of pollution experienced by the children. Plant trees around ballfields and on any thoroughfare separating the school facilities from the transportation pollution from major highways. Create tree classrooms—areas surrounded by trees that will grow to cover and touch creating a tree ceiling.

6. Vote for school board members and policies that educate students in safe and resilient school buildings.

CHAPTER 9

CLIMATE CRISIS IS AN ACCOMPLICE TO THE MURDER OF GEORGE FLOYD—LINKS BETWEEN CLIMATE CHANGE, EXTREME WEATHER AND VIOLENCE

On a warm September night in 2020, I was driving to my office, ready to pour onto paper the thoughts that had gathered throughout the day about climate. During the day I am focused on my job at the Environmental Defense Fund. At night, when I can pull away with a little time, I write. On this particular night, I was driving home deep in thought about Hurricane Sally striking the coast of Alabama and Mississippi. This was the second hurricane to strike the Gulf Coast in less than a month and my mind quickly filled with words I planned to write on storm devastation and the invisible yet lingering impacts of air pollution and flood waters. I brought in ideas that intrigued me from a series of articles that connected global warming to weather-related disasters to an increase in violence. I was determined to avoid all distractions no matter what to get these thoughts on paper.

Up ahead on the road I saw blue flashing lights: two police cars were behind a late model Nissan Altima. As I slowed my car to pass, my heartbeat sped up—although it was dark, my headlights shined upon two young Black men who looked like college kids. School at the University of Mississippi had recently started back with limited in-person classes and these young men were wearing the classic Ole Miss "uniform" of khaki pants and a polo style shirt. They sat on the hood of one of the police cars, heads down, while one officer with a flashlight looked at paperwork and the other looked through their car.

I took a deep breath.

The police brutality targeted at young Black men had reached a palatable all-time high. The June 2020 *Time* magazine featured the depiction of a Black mother holding a space where a child should have been. In the border were the names of thirty-five Black Americans killed by the police. March after march, protest after protest, speech after speech, African Americans have tirelessly worked to bring attention to the injustice hitting our homes and families but it keeps happening. I am the mother of two Black sons, one who has had his own experience with the police. It had been less than six months since the murder of George Floyd. My husband's oldest son, my bonus child, was living in Brunswick, Georgia when Ahmaud Arbery was shot and killed by two white men. The hashtag #iwillstaywithyourson had been trending on Facebook. Mothers of all backgrounds from across the country were taking a pledge to stop when they see young Black men stopped by the police. The charge is simple—take no action, just stay. Just be the presence. Be the mother. Be what you want someone else to be for your son. After the hashtag started, dozens of stories popped up of people either stopping when they saw a police stop or young Black men sharing their gratitude knowing that someone stopped to make sure they were okay. Climate writing be damned, I was going back to sit and make sure those boys were okay.

Before I knew it, I'd pulled into the exit ramp to loop around and dialed my husband.

"Babe, I'm on Highway 6 and the police have two young Black men pulled over and sitting on the side of the road. I gotta turn around." My husband responded with alarm. "Where are you? What highway? What's happening?" I told him my exact location and he could hear in my voice my determined concern. I could hear the fear in his. I knew that there was a chance I might get a ticket or worse yet, be harmed for trying to help. I also knew that violent interactions are more likely to take place when it's hot. When I saw those two boys, I could not help but see my own two sons. The deafening collision of movements that I experienced that night reminded me of a stark reality. Regardless of whether it is climate change or police brutality, there is no social justice issue that does not intersect for Black Americans.

THE BLOCK IS HOT

Extreme heat is one of the effects of global warming that has multiple dynamics in how and where it impacts communities. Unlike a hurricane, flood, wildfire or winter storm, the rising temperature in an area slowly increases and then decreases throughout the day. But since 1901, average temperatures in the United States have risen steadily and, beginning in the 70s, at an alarming rate. According to a report from Climate Central, temperatures in Minneapolis, where George Floyd was murdered by police, have risen 2.3 degrees Fahrenheit since the 1970s during the day and more than 4.3 degrees at night. The number of above-average hot days is up by 25 percent over the same period. Temperatures are cooler in the mornings and evenings, but the midday and afternoon sun can be brutal. Dependent upon the region, the humidity can either create a dry heat that feels warm one minute then burns your skin the next, or a wet heat that makes your hair turn from a sleek, fresh press and curl to a full-blown Angela Davis afro in a matter of seconds. It creates a false sense of day-to-day normalcy—when listening to the weather, rarely does one take into account the average highs and lows and its relationship to how we experience the elements.

But one of the biggest disparities in how we experience heat has to do with where we live. Racist redlining policies of the 1950s and 60s created Black neighborhoods that were unprotected from pollution, waste and unregulated sprawl. Urban communities lacking green spaces and surrounded by concrete, asphalt and gravel experience higher temperatures than suburban and rural areas. The phenomenon is so obvious that there is a term associated: urban heat islands. Urban heat islands are neighborhoods in metropolitan areas with high concentrations of impervious surfaces that hold heat. Black communities that were formerly redlined throughout urban America—Baltimore, Maryland; Chicago, Illinois; Detroit, Michigan; Richmond, Virginia; Philadelphia, Pennsylvania—all have neighborhood heat island communities. The highways that split neighborhoods accompanied by parking lots, asphalt, concrete buildings and sidewalks, and transportation pollution, all of these elements make the neighborhood hotter. In some cases, there can be a temperature difference of 22 degrees between an urban neighborhood and a suburb less than a mile away. The former redlined neighborhoods where George Floyd lived in Minneapolis are almost 11 degrees warmer than surrounding suburban communities. The result is higher energy costs, increased susceptibility to air pollution and heat-related illness and unfortunately in some cases, death for elderly people who reside in these areas.

THE EXCUSED INTENSITY OF HEAT-RELATED VIOLENCE

The Tulsa Massacre, an evisceration of an entire vibrant Black community by a white mob, took place in June 1921. The murder of fourteen-year-old Emmett Till at the hands of an emboldened white mob took place in August. Andrew Goodman, James Chaney and Michael Schwerner, three civil rights workers in Mississippi, were killed by white men in June.

Although science and data back it up, we don't need the numbers to tell us what we know from lived experience. Violence is more likely to occur in hot weather. When layered with years of racial oppression, discrimination and the unofficial sanctioning to police Black people, the warmer it gets, the more susceptible Black people become to hostility and bloodshed.

In 2013, the *New York Times* printed an article recounting the study and connection between weather and violence. After reviewing over sixty of the world's top science journals and data sets, they concluded that the connection between an increasingly warm planet and increasingly violent conflicts is irrefutable. "Decision makers must show an understanding that climate can fundamentally shape social interactions, that these effects are already observable in today's world and that climate's effects on violence are likely to grow in the absence of concerted action. Our leadership must call for new and creative policy reforms designed to tackle the challenge of adapting to the sorts of climate conditions that breed conflict." More recent studies show consistently that violent crimes increase during the summer and the warmer the weather, the more aggressive the behavior. A Brazilian study of "climate change and crime in cities" identified two important factors that contribute to an increase of violence and are found in former redlined neighborhoods and low-income communities throughout America: pollution and extreme heat. These two factors were the sum of evidence that there will be a substantial increase of crime among these areas as we experience climate change. In a follow up 2018 piece by the *New York Times*, the authors charted the increase in violent gun crimes based upon the temperature of the day. In cities like Chicago and New York, twice as many people were shot when the weather was hot versus cold. In the South, the trend was similar but not as pronounced. When I served as Mayor of Greenville, Mississippi, my police chief would always warn me: "Get ready, summer is coming, and folks are about to start acting foolish." He wasn't far off base. The social and psychological effects of heat include aggression, irritability and frustration. Summer heat brought along with it more sexual assault cases, domestic violence accounts and, sadly, more child abuse. If there are identifiable indicators of extreme heat on human behavior, wouldn't that equate to heightened aggression on both sides? We cannot ignore the climate-related factors associated with police brutality that go unnoticed. It was 84 degrees—in Wisconsin—when Jacob Blake was shot by police. It was 61 degrees in Minneapolis on the day that George Floyd died, but people fail to take into account that he died with his face pressed to the street—hot asphalt, an impervious surface that not only retains heat, but gets hotter throughout the day. Rarely do we account for the impact of race, the policing of Black people and the history of discrimination in our country as part of the equation for the intensity of violence of law enforcement against Black people.

The weaponization of calling the police to regulate the "unapproved" social actions of Black people by white people has become more visible thanks to technology and cell phones. Countless videos and 911 call recordings of white people calling the police for some mundane regular everyday action created an entirely new frame of frustration because everyone could now visibly see the privilege white people had been enjoying and what Black people have been complaining about for years. It is especially aggravating and discriminatory when the very event is outside, in nature or environments that created a natural escape from climate heat aggression. The act, whatever it may be, is judged as criminal just because Black people are doing it. If you are white, why worry about the integration of parks when all it takes is a phone call to the local precinct and they will disperse of whatever ruckus concerns you? If you are Black, why go to the park in the first place when you know that by simply jogging along the path, you have to be conscious of your clothing (hoodie = criminal), distance behind a white person (too close = potential criminal) and speed (too fast = runaway criminal)? It creates too much stress. Yes, we need fresh air but damn, do we need a visible permit to breathe?

For some, the answer is yes. "Permit Patty" was the name given to the white lady who called the police on an eight-year-old Black girl because she was too loud as she was selling water on the sidewalk in San Francisco.

"BBQ Becky" was the identity given to Jennifer Schulte, the white woman who called the police on a Black family grilling in a designated area at a park in Oakland, California. Imagine it, a bright sunny day . . . uncles bickering back and forth about what time to put the meat on . . . aunts and cousins unloading coolers of sodas, potato salad and those Little Hug drinks for the kids . . . somebody yelling at somebody's child, "GET AWAY FROM THAT STREET BEFORE YOU GET HIT," as Frankie Beverly and Maze croons from the car radio. Everyone finally thrilled at the sight of a bright sunny day and the opportunity to get outside and away from the stuffy walls of the house and workplace. Then all of a sudden, a white lady is standing close to your picnic set up while calling the police because your family outing has made her afraid. Why do we excuse the heightened intensity of the response related to extreme heat yet created by racism?

The failure to consider the disparate impact and social strain of climate change has created a presumed privilege and acceptance of increased violence by law enforcement and those it empowers, particularly among communities most vulnerable to climate change. The exchange between Amy Cooper and Christian Cooper, who are unrelated, elevated the intersection of Black people in nature, heat aggression and presumed privilege of white people to regulate it, into view for all the world to see. In July 2020, Christian Cooper was enjoying birdwatching in New York's Central Park. Amy Cooper was walking her dog. Christian Cooper asked Amy Cooper to leash the dog, an act required in the Ramble where they were located. Instead of complying with the posted rules of the area, Amy Cooper called 911 and reported Christian Cooper as threatening her. Christian Cooper recorded the entire exchange on his cell phone. The event has been viewed well over forty-five million times. Christian Cooper is a Black man. Amy Cooper is a white woman. Had it not been for the video evidence, every Black man in America knows the incident could have taken a completely different and deadly path. The acceleration of climate change and extreme heat would mean that continuing down such a path would be quite literally a death sentence for people living in urban cities across America. Nevertheless, for the first time in recorded American history, a white person faced charges for wrongfully calling the police on a Black person. We've got to figure out how to protect people, places and the planet without sacrificing the dignity and respect we owe to each other as human beings.

OFFICER CLEMMONS AND MISTER ROGERS

Ever wonder why police uniforms are still dark blue or black?

I mean, think about it. Originally the color of law enforcement uniforms was meant to separate America from the British. Our officers wore blue to distinguish themselves from the British red. But the Revolutionary war is long behind us, and around the world police uniforms have evolved to include other identifiable colors including bright yellow and pale blue. Most police departments in America don a monotone "all black" that intensifies in its solidarity the more violent the encounter. In city after city, rows of officers wearing black tactical gear, black gloves, black tinted helmets and black weapons were the presentation of force against unarmed protestors in the wake of George Floyd's death. Even the police dogs wore black. Despite the uniformity, it begs the question:

Aren't they hot?

That question, among others, was answered in one of the most renowned and groundbreaking episodes of *Mister Rogers' Neighborhood.* In May 1969, the beloved kid's TV show host Fred Rogers addressed the topic of racial injustice that was growing throughout the country. Pool segregation had become the center of civil unrest as well as the use of extreme violence against those protesting its existence. City-owned pools were segregated until the 1964 Civil Rights Act passed and instead of moving forward with desegregation, cities chose to close pools altogether. In St. Augustine, Florida, a hotel owner was captured on film pouring acid into a pool of integrated swimmers. In my hometown of Greenville, Mississippi, the city-run public pool in the historically Black neighborhood was filled with dirt and concrete in the mid-1970s. Since both public high schools had pools, Greenville High School, which was white, and Coleman High School, which was Black, remained open but segregated by geography. By the late 1980s, the pool at Coleman closed and all funds were funneled to the larger, historically white pool at Greenville High School.

In 1969, cities and towns were grappling with integration and heat resulting in violence that often included children. Mister Rogers shared a clear answer on his show that reckoned with not only racial injustice, but dealt with the underlying issue of heat and a remedy we could embrace. He did all this in less than two minutes using a wading pool, water hose and a towel.

In 1969, the message was targeted to Mister Rogers's audience: white middle-class children. That was the demographic that had access to an in-home television, let alone time to watch the show. Mister Rogers was inclusive of Black people, but the message was meant to be received by the base of the power structure in America and to shift a generation. The scene begins with a pair of white feet in a pool being cooled off by water. Mister Rogers is cooling off in a fenced-in yard, trees and greenery in the background, when Officer Clemmons, a Black man clad in a long-sleeved police uniform and jacket, comes to join him. Mister Rogers acknowledges the heat, then offers him a seat to join him in the wading pool. Officer Clemmons is hesitant because he doesn't have a towel. Mister Rogers offers to share his towel. Together they share the pool and cool off together. In less than two minutes, Mister Rogers acknowledged the difficulty of the physical heat, showed awareness that as a white person he too felt the same heat, created an invitation to share in a solution, then removed the barrier of inequity by sharing his resources. It is the equation to reducing climate heat impacts and thereby removing the factor that creates the violence. Acknowledgment + Equity x Resources = Solutions.

The message in 1969 was meant for white middle-class children who would later become part of the majority leadership structure for America.

When leaders didn't get the message and instituted equitable policy changes, Black folks shared their own version. In 1982, *Saturday Night Live* debuted "Mister Robinson's Neighborhood" featuring Eddie Murphy as a parody of *Mister Rogers' Neighborhood*. Funny as hell, the episodes recount the reality of Black urban life while giving a complete opposite of Mister Rogers's equation. Mr. Robinson's neighborhood is the reality of what happens without the acknowledgment of racial injustice while highlighting the disregard for equity and hoarding of resources. It lasted six seasons.

Years later, amid the 2018 continued and unsettled racial unrest, Mister Rogers and Officer Clemmons reunited and recreated the scene, adding layers of familiarity. Instead of the characters, it was François Clemmons and Fred Rogers. François, known as Officer Clemmons, wears a light blue shirt and has shed the heavy police coat in May. Fred wears his classic cardigan in a rich earthy green. The message needs to be reiterated to all of us at this time and we need to put it into action. The violence and extreme temperatures increase at the same rate as climate change. It's still hot and we're still here together.

The night that I pulled over to make sure two young Black men were okay as they were being stopped by the police, I called the police department to identify myself, the description of my car and requested that they please notify the officer on the scene that I was pulling over to observe. As dispatch connected to the officers, a bright light shined through my passenger window. With a brief explanation, the polite white officer looked at me with sincerity and said, "Ma'am, I completely understand and respect why you stopped. The young men are okay, and I'll let them know that someone was making sure they are safe." I breathed a sigh of relief, grateful that God continued to send angels to watch over the craziness I pursued and hopeful that my husband wouldn't fuss too badly about my actions. But if we could use that equation from Mister Rogers and Officer Clemmons—acknowledgment with equity and shared resources—we might have a chance of not only eliminating unnecessary police brutality and violence against Black people, but at living better together while adapting to a warmer climate reality.

MAKE IT MAKE SENSE—CLIMATE AND VIOLENCE TERMINOLOGY

Extreme Heat

The literal meaning of "It's hot as hell." Extreme heat occurs when an area's summertime temperature exceeds the average for more than two to three days in a row. The problem is that as global warming increases so does the average summer heat. Every year the average gets higher and the extreme grows larger. Effects of extreme heat on the body can be deadly. Heat exhaustion, dehydration and worsening of chronic disease are becoming more frequent among older adults, young children and anyone working outdoors in extreme heat.

Impervious Surface

Pavement, concrete, asphalt or other hard water-resistant surfaces that do not allow water or air to flow through them. Impervious surfaces hold water and prevent rainwater from flowing in a natural pattern in the ground. While we want impervious surfaces on the roof to prevent leaks and protect homes, we do not want impervious surfaces at the park or playgrounds holding heat and water.

Urban Heat Island

An urban heat island is created when an area, neighborhood or community is covered with impervious surfaces like concrete, brick and asphalt and little green space. The impervious surface area retains heat, making the area hotter than the surrounding community without as much hard surface. The effect results in warmer temperatures, greater air pollution, less airflow and a greater risk of heat-related illness. Urban heat islands can often be found in low-income communities with little to no green space.

BEFORE THE STREETLIGHTS COME ON

1. Join your local community policing initiative. If you don't have one, create one. Plan for summer and extreme heat resources to remind the community. Cool off, open pools and open fire hydrants.

2. Talk to your police department about lighter clothing for officers in the summertime. Encourage law enforcement on bikes and walking as opposed to idling engines.

3. Abuse and all types of violence increase during warmer months. Child abuse, domestic violence, assault and murder—when temperatures rise so do tempers. Support alternative dispute resolution sessions in both law enforcement and community action centers and schools.

4. Place activities on grass instead of asphalt. Asphalt is a dark, absorptive surface—it holds heat. Look for reflective surfaces—grass, rock, natural landscapes—to help reflect the heat and reduce conflict before it starts.

5. Vote for community-led, community-driven policing initiatives. Proactive learning exchanges are more effective and safer than defensive responses to violence.

CHAPTER 10

VOTING FOR OUR CLIMATE LIVES—CLIMATE VOTING TRENDS AND VOTER SUPPRESSION

THE SCARIEST HORROR FILM EVER DEBUTED ON JANUARY 6, 2021

There's a good reason Black folks don't say "Candyman" five times in a row and it happens to be the same reason we are more likely to vote for climate policy more than any other demographic in America today. There's no need for us to tempt fate or question, "What if?" when we exist in the lived experience of environmental injustice, continue to rebuild despite increasing violent weather patterns and systemic barriers to capital, all while healing from racial trauma. In other words, Black people do not give potentially life-threatening horror the benefit of the doubt—we know better than that. We vote for policies, including climate and environment that are most likely to keep us safe from both the things we can see and the things we cannot.

My dad liked horror movies, my mother did not. As a kid, I wavered between the "acceptable" alien gore films and the "unacceptable" poltergeist stuff. "Run for your life!" was the catchphrase uttered in just about every single movie. In the 1990s, I was a Black teenager growing up in a modern version of Anne Moody's Mississippi, so the reality of the words, "run for your life" were not just fictional fantasy horror, but part of our learned historical reference. It translated into "move your behind or die" and there were real stories and headstones of Black people who did not.

Watching the movie magic of how these words played out for the rest of America was fascinating and thrilling. My friends and I would do our best to get into our town's one theater to see the latest guts and gore slasher movie knowing full well that A. The likelihood of getting into a rated "R" movie on an opening weekend night meant knowing someone that worked at the theater and B. Our parents would not approve, especially of any horror theme invoking demons or the dead. We'd might as well wear signs proclaiming "spirit to be rebuked here" when we entered church service that weekend because surely the church mothers would see the glow of heathen activity all over us. The other alternative was to figure out who had a Blockbuster card with the least amount of late fees, rush to grab the latest video and convince somebody's mama to let us have a sleepover. Either way, we'd gather and stare at the screen in utter disbelief.

Any elevation or animation at the word, "Run!" and the people I grew up with weren't sticking around for the rest of the sentence. It was a foregone conclusion to move as fast as possible in a direction opposite of the drama and ask questions later. But for some crazy reason, white people ran towards the conflict. At some point in every movie, the obvious victims would wander aimlessly into a clear and present danger, oblivious to all signs and signals warning them to go the other way. These actions made no sense, even in our fifteen and sixteen-year-old brains. Why would you go against your own safety and interest? What level of curiosity overwhelmed self-preservation and good old-fashioned common sense? Despite our yelling at the clueless characters on the screen, we shook our heads while clutching each other and hoping desperately that they would see the error of their ways and run like hell back to a place of good common sense.

It never happened.

We knew the ending within the first five minutes of the movie. The threat was almost always slow-moving and, at the pivotal moment of awareness and escape, we'd all throw our hands up and yell, "Awww come on!" when the victim fell and couldn't get up. Every character in the movie seemed to be in the dark, both literally and figuratively, about the danger. We knew who would die and who would live to fight another day in the obvious sequel. No matter the setting, plot, character outline or story arc, the traumatic theme we gleaned was always the same. If you blatantly ignore the warning signs of danger, you will suffer the consequences. At the end of the movie, we'd joke about the one commonality in all of these films: white women were always counted among the victims. I shook my head at the ignorance of the victims. Would they ever learn?

On September 29, 2020, we watched a trailer for the real-life horror film of America in the form of the first presidential debate. The movie posters were unveiled through a bitterly cruel election cycle and period of racial unrest throughout the country. After a close November 2020 election, the flip of several Republican-held states and the Georgia run-off election victory of Rev. Ralph Warnock on January 5th, we all watched the movie premiere live at the United States Capitol when those opposed to the election turnout violently lodged an insurrection on January 6, 2021.

Blood spilled.

Bones were broken.

Flesh was crushed.

People died.

The 2021 horror film of the year could aptly be titled, "American Vote of Death" as our democratic process of selecting leaders was fraught with shock, dread and fear. Voting—the process, access, who participated, who counted—teetered on becoming a terrifying urban legend and, to this day, some question whether or not the election was real. At the same time, the climate crisis continues to loom larger and more impactful by the day making the need for government climate policy an urgent requirement.

Just like those 90s horror films, the present danger of suppressing and devaluing the process of voting was obvious to many Black and brown voters early in the 2020 election cycle, yet considered a "what if" for white moderates. The real threat of a climate crisis combined with systemic oppression of minority and marginalized people was given the benefit of the doubt by Donald Trump and Mike Pence during the presidential debates. They lured voters into thinking that climate change wasn't that bad, justice abounded for all and paraded their one or two Black supporters across every network to prove it. After the elections, the signs continued to scream, DANGER AHEAD, when states that Donald Trump lost began recounting, trying to throw out votes and even threatening their own members. After losing both Senate seats in Georgia and later three recounts, the Georgia Secretary of State, Ben Raffensperger, was publicly pressured by the President of the United States of America, to change the numbers of the recount. When he ignored the calls of Donald Trump (taped voicemail messages from the President himself) and declined to alter the process because it was illegal, he suffered threats to his future political career and bodily harm to his wife and children. It was like the 1990 film version of *Night of the Living Dead*—after realizing their community had all turned into zombies, the survivors thought to seek shelter deeper in the bowels of their house but ended up being eaten by their own.

By December 2020, state legislatures were planning to restrict and revise voting procedures across the country and by May 2021 over 361 voter restrictive bills had been introduced in forty-seven states. By July 2021, seventeen states had passed at least one. The bulk of the American public granted a benefit of the doubt and thought that once President Biden was sworn into office, things would begin to normalize and settle. Surely those in stark opposition would have to accept the election results, right?

The open vitriol of the 2020 election cycle, aggravated by the antics of a sitting president, combined with the COVID-19 pandemic, raw feelings from the deaths of George Floyd, Breonna Taylor and Ahmaud Aubrey created a palatable difference. This was not going to end well and Black people knew it early. Black, brown and Indigenous organizers, along with many allies, vigilantly focused on protecting and turning out the vote for special elections as well as continuing strong voter registration efforts. Stacey Abrams, former Georgia gubernatorial candidate and the first Black woman in Georgia to secure a major party nomination, was often touted as an automatic nominee to any part of the Biden administration she wished to serve in. She went on ABC's *The View* to not only dispel the rumors but to stress the importance of voting and the fact that the struggle against voter suppression was real and ongoing in Georgia. She could not leave at a time like this. The voter suppression impact was not lost on the environmental community either. Environmental justice organizations sprang into action by connecting the dots between voter suppression and environmental justice for voters in impacted communities. Almost every major green group supported efforts to ensure that the simple right to vote was not hindered by partisan political influence but instead allowed for equitable and easy participation for all. Climate policy was at risk and voting rights were not pursued collectively by the entire environmental community. Federal climate policy, let alone equitable federal climate policy, would be lost.

At an almost perfect duplication of the horror movie, it wasn't until it was too late and on January 6th, we watched our nation's capital fall under a domestic attack based on a voting myth. When images of the southern confederate flag loomed across screens around the world while members of Congress sat huddled in safe rooms and the Vice President of the United States of America was escaping for his life, that's the moment America heard, "Vote for your life!" Meanwhile, Black people heard, "Vote!" months ago and had already started registering, advocating, preparing and knowing full well the scenes that lie ahead.

FLASHING SIGNS OF DANGER AHEAD

Amid a pandemic made worse by the impacts of environmental injustices on marginalized communities, extreme weather exacerbated by the effects of climate change and years of structural racism uncovered day after day, voting is one of the few surefire ways that Black people can participate and have a major impact on policy. Study after study has shown that Black and brown people are strained the most by climate change. The people most impacted by the threats of climate change are the same people most likely to vote for climate pollution control policy but are being suppressed in their access to the polls.

Nevertheless, watching the 2020 Presidential and Vice Presidential debates made me clutch my proverbial pearls. Climate change got a shocking eleven minutes of discussion for the first debate. No one expected it, and debate moderator Chris Wallace didn't list it as one of the topics to be covered beforehand. After listening to countless minutes of toddler-like bickering, Chris Wallace directed the question to Mr. Trump, "What do you believe about the science of climate change, sir? Do you believe that human pollution and greenhouse gas emissions contribute to climate change?" It was the unexpected plot twist in the horror show. Had we screamed and shouted loud enough that climate change was finally being received and respected for the crisis it really is? I was stunned but glued to the screen for more.

Reality set in as the world heard the President of the United States insinuate that forest management, also known as raking the forest floor, would be the single most effective prevention of global warming. He claimed that he wanted "immaculate air and water" but refused to address the more than one hundred rollbacks to Environmental Protection Agency (EPA) regulations that will ultimately increase the pollution we experience, particularly in Black and brown communities. According to Mr. Trump, the Paris Climate Accord was bad for business and relaxing the fuel economy standard was just a "tiny bit" of an impact. In later debates, the world listened to his claims that renewable energy was bad because the windmills killed all the birds, environmental justice communities were fine because they were employed (by the fossil fuel industry) and making more money than ever, and that everything will be solved because he's planting a trillion trees. This was the same man that claimed water efficiency had to do with low shower pressures and people with runny toilets having to flush them multiple times. Immediately I thought, "Who's listening to this? Who is making a decision based on this conversation? Who thinks Trump and Pence are right?" If climate change was not directly correlated to other factors driving voters to turn out—things like justice, healthcare and jobs—then the future of climate policy and climate safety was at grave risk. Not voting was not an option. This was a vote for our climate lives.

The vice presidential debate was its own horror flick, complete with guest appearances by an insect reminiscent of the terrifying scenes from Jeff Goldblum's 1986 horror thriller, *The Fly*. Who among us didn't recognize the feeble attempts of bullying that Vice President Mike Pence tried when he attempted to speak over Senator Kamala Harris? Bless his heart; he tried it. But even his mansplaining and privilege could not negate the fact that she talked about climate and environmental policy in a way that spoke directly to the needs of the American people. Calmly but firmly, she addressed issues of environmental justice and how America must meet the climate crisis head-on by re-entering the Paris Climate Accord, investing in renewable energy and creating clean energy jobs. On the other hand, Mike Pence refused to acknowledge climate change as an existential threat and had the audacity to claim we were experiencing the same number of hurricanes that we did one hundred years ago on the eve of Hurricane Delta hitting the southern United States. I wanted to scream as loudly as I did at those horror films: "WE'VE RUN OUT OF NAMES, MIKE. We've run out of names for the hurricanes. Don't you think this might be more intense?"

Despite the Trump/Pence debate strategy to disregard, devalue and banter the Biden/Harris team, at least the latter came with a plan. Before the 2020 elections, environmental justice had never entered the debate platform once, let alone all three events. Rising emissions, loosening regulations on the fossil fuel industry that exacerbated health problems for citizens, increasingly violent weather events and a vocal and present youth movement all elevated climate change as a real voter issue. The third and final debate was barely scheduled due to COVID-19 exposure. It was initially canceled due to the president's coronavirus diagnosis and failure to accommodate for the safety of all debate participants.

Environmental leaders knew that the science was real, but the stronger political argument were the climate and environment effects that could be personified and felt. To characterize it otherwise was a wholesale effort to do more than just perpetuate a false myth about climate, it was an attack on the ideology of voting. If there's no real issue, then there's nothing to vote on right? That is exactly what the Trump/Pence team hoped would happen and it did. The entire exchange was that familiar scene in the horror film: someone with the plan of escape is punted down by the idiot that thinks nothing is wrong.

STOP TRIPPING OVER YOUR OWN FEET AND RUN!

Moms Mabley, one of the first Black female comedians in America, had a funny way of saying a phrase common in the Black community: "If you always do what you always did, you will always get what you always got." In 1980, Audre Lorde, a Black lesbian feminist author, detailed exactly how seemingly shared experience of womanhood plays out between Black and white women in America.

> Today, with the defeat of the ERA, the tightening economy and increased conservatism, it is easier once again for white women to believe the dangerous fantasy that if you are good enough, pretty enough, sweet enough, quiet enough, teach the children to behave, hate the right people and marry the right men, then you will be allowed to coexist with the patricarchy in relative peace, at least until a man needs your job or the neighborhood rapist happens along...Some problems we share as women, some we do not. You fear your children will grow up to join the patriarchy and testify against you; we fear our children will be dragged from a car and shot down in the street, and you will turn your backs upon the reason they are dying.

Over thirty-five years later, the humor of Moms Mabley and the honesty of Audre Lorde combined to produce the horror of November 2016.

In 2016, only 45 percent of white women voted for Hillary Clinton versus 98 percent of Black women. Despite having held public office, being arguably the most qualified candidate, and certainly the one who didn't molest or illicit open sexual violence toward women, white women were split on casting their vote with the one person more likely to support climate policy and equity. On the other hand, Black women were unequivocally unified; 98 percent voted for Clinton. It was like the beginning of *Friday the 13th Part VII* in 1988 when the white women in the movie were responsible for releasing Jason from the bottom of a lake to go on another killing spree. In our heads, we all screamed, "WHY WHITE WOMEN? WHAT IS YOU DOING?" Black women felt like we watched our sisters walk through the deadly forest of election rhetoric knowing full well that the boogeyman would jump out. They heard the buzz of his knives and saws in "hot mic" moments and "grab them in the..." comments. Despite it all, many walked into the nightmare willingly. The 2020 election cycle provided a better ending sequel; however, the post-election analysis concluded that lessons from the 2016 loss were only moderately learned. While a lower number of white women voted for Trump, he surprisingly picked up more Latino women, a group that typically votes highly in favor of equitable climate policy. The saving grace in 2020 was the record voter turnout—almost a 7 percent increase in voter participation. But the increase and apparent "win" for voter inclusion became the key theme for Trump republicans and far right-wing extremists in perpetuating the "myth" that the 2020 presidential election vote was not valid, and the entire thing was rigged from the beginning. The result has been immediate and the looming state challenges have promised to be just as fierce.

Instead of concentrating on policies that would protect citizens from the immediate impacts of the climate crisis, some states directed attention to reducing the number of particular voices participating in the process. This is true in areas where Black and brown people show a stronger affinity in voting for climate policy than their white counterparts. Across the country, Republican held state legislatures and governors are laser-focused on limiting access to voting by redefining who is eligible, creating additional measures of proof of citizenship and then adding backend stop gaps that will allow say, a Republican secretary of state, governor or Republican held state legislature, to overrule the popular vote. Shocked at the loss in the typically right-leaning state of Arizona, the state legislature presented and passed bills to limit early voting, requiring more stringent signature matching and went so far as to strip the Democratic secretary of state of some election powers. While the state focused on passing restrictive voter laws, citizens in the state suffered from water shortages and wildfire spread—all issues related to the climate crisis. In Georgia, possibly the most contested swing state in the nation, state legislators gathered and passed legislation to limit mail-in voting and increased voter ID requirements. They tried to eliminate Sunday voting, also known as "Souls to the Polls" day for many organizers in the faith-based community. Fortunately, that measure did not pass but brought great awareness to the political hypocrisy of the religious right. They could not justify it being okay to attend the Sunday football games at Falcons stadium while claiming that Jesus would not be pleased if you voted on your way to the stadium.

LESSONS FROM LOVECRAFT COUNTRY

Over the years, things have changed in the horror genre. Shows and movies like Jordan Peele's *Get Out* and the updated 2021 version of *Candyman* are exploring the racial and justice traumas that produce real and relevant fears and anxieties among the Black American community. But none has been more applicable and in your face than the HBO series, *Lovecraft Country*—an American horror series that reveals the terrifying trauma associated with racism while reclaiming the power of our own story. Set in the 1950s, *Lovecraft Country* explored the intersectionality of Black trauma from racism, sexual identity, voter suppression, housing discrimination, healthcare, education and climate science—all wrapped in a familiar need passed from generation to generation. The terror of American racism is real, give no benefit of the doubt and do whatever it takes to survive. *Lovecraft* is so crosscutting that it has a soundtrack, podcast and an official syllabus. I was devastated when it was not renewed for a second season. I see elements of *Lovecraft* lessons played out in the fight to secure voting rights to have a voice at the table for an equitable climate future.

When Texas Governor Abbott demanded all state legislators return to the Texas state capital to pass the Voter Integrity Protection Act—a bill that would restrict access to thousands of minority voters throughout the state—Democrats refused to return and left the state. Under penalty of law and arrest, fifty-nine members of the Texas Democratic Party left their families, boarded flights and arrived in Washington, DC to advocate for voting rights for citizens in Texas and around the world. For constituents who live under a constant cloud of oil and gas influence and pollution in metro areas, strengthening hurricanes and floods along the coast and an electrical grid that not only shut down in one of the worst winter storms to hit the state but also highlights the inequity of energy cost to poor communities, having a voice through their vote is a lifeline.

In Georgia, lawmakers sought to end Sunday voting and limited the items that could be passed to voters standing in line. They went so far as to say one could not give water to people standing in line in July. The pushback was fast and organic. I watched a TikTok video of an older Black lady unloading her trunk at a voting precinct with a setup that rivaled a SWAC football tailgate: popup lawn chairs, water, chips, foil-covered plates of food, cell phone battery chargers, fans and a radio. Her message was simple. "WE (Black people in Georgia) are not new to this, we are TRUE to this. We are the offspring of Dr. Lowery and Dr. King. Try us if you want, we're already ready." These homegrown messages spread through every form of social media available. Tactics meant for suppression had the opposite effect and people registered to vote and turned out for the January 2021 runoff election en masse. The battle between voter suppression and equitable power took on the appearance of good versus evil and on January 5th, the day before the US Capitol riots, Georgia voters elected Rev. Warnock in a close election.

Like the powerhouse leading women of *Lovecraft Country*, more women of color are not only running for office, leading movements and preparing people to turn out to vote in every election, but they are centering climate change and equity as part of the work. The *Time* 100 list of most influential people for 2020 had more women than ever and included two remarkable climate justice female leaders—Dr. Ayanna Johnson and Cecilia Martinez. During the 2020 election cycle, Facebook groups like "Suburban Women for Harris" quickly grew to more than 200,000 active members and TikTok videos of white, suburban, climate moms waging and winning savage sign wars with their less than liberal neighbors garnered hundreds of thousands of views. But the fight is far from over and the climate crisis remains a major factor impacting our homes, families and lives. We cannot sit idly by and watch the voting rights of millions disappear while those intent on holding power use unethical and diversionary tactics to sway the rest of us. Black people have influence and voice that must be broadcast louder. If not, we are ignoring the warning signs. I've heard often the phrase, "We're tired of saving white women," and to some extent, I agree. But what we cannot do is sit on the side of the road whistling and waving as we save our own but watch them and others walk into the dangerous dark forest with all signs pointing to impending doom and destruction. Because the truth is, trauma is truth, horrors are based in reality and zombies do exist in real life. Instead of ignoring the zombies that can come back to eat us alive, it's better to burn the whole damn forest to the ground. Let's not settle for the same old *Friday the 13th* or *I Know What You Did Last Summer* ending. This is *Lovecraft Country*—Leti, Ruby and Hippolyta fierceness. Strategize like Ruby, know the numbers and science like Hippolyta and be as fierce and bold as Leti. It's time out for letting voters walk into that forest. Now is the time to block the path and lead the way in the opposite direction, expecting everyone to vote for all our lives.

MAKE IT MAKE SENSE—CLIMATE AND VOTING TERMINOLOGY

Voter Suppression

Voter suppression is any tactic deployed to discourage, limit, delay or otherwise block access to voting for any specific group of people. Article 1 of the United States Constitution grants all US citizens age eighteen and older the right to vote. The 15th Amendment protects the right by declaring the right to vote shall not be denied on the basis of race, color or whether one was previously enslaved. The reason this amendment was necessary was due to the individual state establishment of laws that prevented Black Americans from voting. Not to be outdone, states became creative. They instituted extensive voter registration applications, outlandish literary test and fees called poll taxes and restricted access to voting based on whether one could pay or not. States then began changing the political boundaries and lines so that the outcomes of voting would be determined by where you lived, an attempt to use segregation to influence and suppress Black voting in America.

By 1957, the United States passed additional voting rights legislation to quell voter suppression and keep an eye on the worst actors—southern states. But in 2013, the United States Supreme Court changed the rules by shifting the burden of proving voter disenfranchisement from the state to, you guessed it, the voter. The result was immediate as states, particularly in the south where large populations of minority voters live, began reinstituting voter discouragement tactics without fear of retribution. By 2021, voter suppression laws included everything from preventing people from sharing food and water while standing in a long line to vote (Georgia, Senate Bill 202) to putting election officials in jail as criminals if they failed to remove people from the voting rolls as directed by the state (Iowa, Senate Bill 413).

Voter Restrictive Legislation

Voter restrictive legislation are proposals and bills to enact state laws that limit who and how people can vote beyond federal requirements. Additional barriers such as strict voter ID laws, limits to mail-in voting, allocation of voting machines and voter roll maintenance all make it more difficult for people to vote. These are not requirements by federal law but added by state legislatures to control the voting process. It's the reason you must re-register to vote every time you change your address from state to state. Minority people are more likely to vote for climate and clean energy policy than any other demographic in the country. These are the same people disproportionately impacted by voter restrictive legislation. The people most impacted by climate change are being kept from voting for protection from climate change.

BEFORE THE STREETLIGHTS COME ON

1. Confirm your voter registration at your current address. Do this annually to ensure no changes have been made that affect your ability to vote.
2. Obtain a list of all city, county, state and federal election dates in your community along

with any deadlines to register to vote for the election. This information should be easily obtained from your Secretary of State's office.

3. Elections are held for people AND policies. Research any ballot initiatives that are up for vote. Support policies that provide greater access to voting, create renewable energy opportunities, identify greenhouse gas reduction goals, and support climate adaption and resiliency measures across the community or state.
4. Consider running for an elected office or seeking an appointed political position in your local community or state. Consider this a formal ask.
5. VOTE!

CHAPTER 11

THE ANSWERED PRAYERS OF OUR ANCESTORS—FAITH AND CLIMATE ACTION

The Lord is my shepherd; I shall not want. He maketh me to lie down in green pastures; he leadeth me beside the still waters. He restoreth my soul: he leadeth me in the paths of righteousness for his name's sake. Yea, though I walk through the valley of the shadow of death, I will fear no evil: for thou art with me; thy rod and thy staff they comfort me. Thou preparest a table before me in the presence of mine enemies: thou anointest my head with oil; my cup runneth over. Surely goodness and mercy shall follow me all the days of my life: and I will dwell in the house of the Lord forever.

—Psalm 23 KJV

You, Lord, are my shepherd. I will never be in need. You let me rest in fields of green grass. You lead me to streams of peaceful water, and you refresh my life. You are true to your name, and you lead me along the right paths. I may walk through valleys as dark as death, but I won't be afraid. You are with me, and your shepherd's rod makes me feel safe. You treat me to a feast, while my enemies watch.

You honor me as your guest, and you fill my cup until it overflows. Your kindness and love will always be with me each and every day of my life, and I will live forever in your house, Lord.

—Psalm 23 CEB

I must have learned the 23rd Psalm, also known as "David's Prayer," somewhere between Sunday school lessons, bedtime prayers and vacation bible school. It was a given: every good Christian child could recite the 23rd Psalm. "The Lord is my shepherd, I shall not want." In other words, with God at the helm, I do not and should not need anything but Jesus. The words weren't just recited, we prayed it, sang it and danced it. My favorite musical interpretation of the Psalm was a rendition of "Lead Me, Guide Me," sung by my Aunt Ceola. Ceola Thomas was not an aunt by blood, but one of my mom's best friends and therefore in the tradition of the Black community, my aunt. She was the epitome of politeness, grace and kindness. After a sleepover with her daughter and my "God sister" Donna, she'd wrangle us out of bed, prepare breakfast, fuss over our clothes and shoo us off to St. Peter Missionary Baptist church. Aunt Ceola had a small frame, light airy voice and delightful smile. She could belt out a deep alto melody that would lull any living creature into peace and praise. On any given Sunday morning, I could hear the rich tones of her voice begin in a hum, rising to words, "Lord let me walk, each day with thee." These words were a balm to the soul after a week of rejection and crisis brought on from discrimination, poverty and oppression. My Aunt Ceola has since passed away but I will never forget how I was instructed to embrace the words of the 23rd Psalm as both a sense of intimately belonging to God and a warning: greed and excess should be non-existent in my life because God covers all.

It was the description of the scripture that always caused my mind to wander, "He makes me lie down in green pastures. He leads me beside still waters. He restores my soul." I imagined myself lying in meadows of green grass beside mirror-like lakes of water feeling refreshed, relaxed and at peace. Never mind that I'd never actually seen a green pasture—most pastures I knew of in the Mississippi Delta were full of cows, horses and other things that one wouldn't want to lie down in. The waters I knew were those of the mighty Mississippi River. Common sense told me that it wasn't wise to be led to or stand anywhere close to the river's edge. People had been swept away, never to be seen again for doing much less.

The imagery of nature reflected as peace inundated every semblance of scripture I knew. Listening to the choirs at St. Peter or my home church, Agape Storge Christian Center, sing of the rivers of Jordan, never created an image of polluted streams and ditches that ran alongside the neighborhood. I imagined David's heavenly river surrounded by expansive gardens and shining clear skies. A lush green environment that replenished the body as well as the soul. I don't recall reading a single Sunday school lesson about hurricanes in heaven. On the other hand, floods pestilence and plague? All associated with a sinful world hell-bent on making the earth "groan" as described in the book of Romans. The peaceful environment embodies freedom, something that Black people have been fighting for in word, song and deed, especially in the Black church. How, when and where would we experience this environmental ecstasy here on earth?

THE EARTH IS THE LORD'S AND THE FULNESS THEREOF

The Earth is the Lord's and the fulness thereof; the world, and they that dwell therein. For he hath founded it upon the seas, and established it upon the floods.

—Psalm 24:1–2 KJV

There is no escaping the relationship between faith, environment and climate action. In the Black church, the images of heaven we sing about, pray for and hope are soon to come are filled with bright skies, clear water and land so full of natural resources that the diamonds and precious stones that we dig up here on earth are found lining the streets of heaven. This is the place that we believe our loved ones go and await our arrival. This is where Jesus has gone to prepare a place for us according to scripture. It is a place that is filled with all of the good things that we don't have enough of here on earth without the pain, inequity and despair. Enslaved Africans were encouraged to envision it as an escape from the brutality of slavery. Many used their knowledge of the natural environment to help them get to heaven either by traversing rivers and streams or to die trying. One way or another, they'd get there.

Somehow, we American Christians became convinced that you can't believe in Jesus and climate change at the same time, which is crazy to me because the actual tenets of the Christian faith have to do with taking care of God's creation. The mainstream environmental movement is mostly white and agnostic making it difficult for people of faith, particularly Black people of the Christian faith, to see themselves in the movement. Even I've admitted that I never saw myself as an environmentalist. Those were the vegetarian white people that hugged trees, saved turtles and wore Jesus sandals but didn't believe in him. That does not negate my responsibility as a Christian believer to take care of what God has blessed me with and part of that is care for the earth. Christianity does a disservice to the environmental movement in couching our Christianity as a "Dominion-ism"—God has given us dominion over these things to do with as we please as opposed to "Creation Care"—God has charged us with taking care of this.

Creation care has a deeper meaning for Black Christians because it reaches back to our enslavement and sets purpose for our salvation. The idea of salvation is freedom, which to an enslaved person is important. To be set free from the physical bonds of enslavement is akin to being set free from sin. Both are considered weights on our humanity. But with freedom comes responsibility—a responsibility to maintain freedom and to free others, to be responsible stewards of freedom while caring better for each other.

During the periods of enslavement, the Civil War and Reconstruction, Christianity meant spiritual deliverance and some semblance of equality for Black Americans even when it didn't exist in their day-to-day lives here on earth. Equality in nature wasn't a question. "That you may be children of your Father in heaven. He causes his sun to rise on the evil and the good, and sends rain on the righteous and the unrighteous" (Matthew 5:45 NIV). The sun didn't distinguish color when beaming down the mid-day heat, trees didn't decide who to protect and not based on one's social status and rivers certainly didn't discriminate based on who was flowing or running through one's current. Some would come to believe that the elements of nature worked on our behalf as a source of protection and that the pigment in our skin more closely aligned with that of the environment. The book of Genesis even describes it: "And the Lord God formed Man of the dust of the ground, and breathed into his nostrils the breath of life; and man became a living soul" (Genesis 2:7 KJV). Nature is the great equalizer. Our ancestors were brought to this continent and suffered insurmountable pain. Some found solace in a newfound faith that was pressed upon them. They grabbed hold of those tenants that were familiar to home—rivers flowing like the Jordan and heavenly blue skies. Maybe, just maybe, part of that faith was a belief that despite physical bondage they could be free through Christ. But there is also an obligation to turn our captivity into freedom for someone or something else. We must take care of what God has given us—access to the land, freedom to move as we will, care of the natural resources and use all of these to free other people—not through our Bible-thumping but through how we live and care for God's creation.

As I teach my children David's prayer, it is difficult to show them the imagery of still waters. Where do Black children leaving Sunday school in Detroit see a green pasture? Green spaces in Black communities are few and far between and harder to create without adequate funding and a desire to see these areas as vital to community development. Once during a work visit of a neighborhood in West Atlanta, residents and members of the West Atlanta Water Shed Alliance (WAWA) took my EPA staff and

I to a community green space that had been developed in coordination with EPA, local community organizations and churches. While on the bus, someone mentioned that a church in Buckhead, a traditionally affluent white part of Atlanta, wanted to partner with churches in West Atlanta to plant community gardens to teach the community about growing healthy food and land use. The Black parishioners in our van scoffed. "We don't need their help. How are they gonna come to our neighborhood and teach us what our ancestors taught them?" I laughed. She was absolutely right. They didn't need some big wig church coming to the neighborhood with Home Depot boxes, seeds and volunteers doing good for the weekend. They needed and wanted access to the same opportunity to create the green pastures in their own neighborhoods with their own hands. They discussed getting access to burned-out homes and overgrown lots: real transfer of land opportunities that could be used to elevate the viability of the neighborhood while teaching the community about land preparation, growing, sustainability and resiliency. They needed access to workforce development grants that would pay local community members to learn trades. It would not only help people better themselves and provide economic stability for the family but also contribute to the neighborhood in a way that would increase property values for homes. They drove us to a park. At one time it had been an abandoned property ignored by absentee landlords, now it was turned into a central location area for children to play while learning about the ecosystem outside their front door. These were the green pastures and still waters David sang about and it could be experienced right here in West Atlanta if we only gave it a chance.

To better understand how influential faith is to understanding how African Americans think and act on climate, we must look at facts surrounding the African American community and religious views. Almost 80 percent of African Americans identify as religious, more than any other demographic in the United States. About half of that number consider themselves to be protestant and historically so. It's in the family. Families attend the same church for generations. Children may leave home to attend college but will find the same type or denomination of church that they attended at home. African Americans that may consider themselves to not be actively religious will observe certain traditions and holidays out of respect for their elders. We affectionately call these people members of a special "CME" denomination. They go to church on Christmas, Mother's Day and Easter.

The largest and most influential Black denominations can be boiled down to the big three: The National Baptist Convention, easily the largest with over seven million members across the world, The Church of God in Christ (COGIC) with a membership of over six million and the largest Black Pentecostal group and the AME (African Methodist Episcopal) church with over 2.5 million members and seven thousand congregations across America. All three were founded in the South during the time of reconstruction. During this period, there was a longing for equity that had been described spiritually but had not yet manifested physically. While churches of each denomination spread to the four corners of the country and beyond, the majority of churches grew and strengthened in the places that boasted strong Black populations yet experienced the greatest injustices and inequities. The conservative South remains a stronghold for Black churches while urban city centers like Detroit, Chicago and Philadelphia are home to large congregations stemming from the traditions carried through Black migration to the north.

African Americans take far more religious actions than our white and Latino counterparts. Despite the images of white evangelical Christians that populate our timelines and take over newsfeeds as the most religious group, it's Black folks that truly fit the bill of being the most religious demographic. Almost 75 percent of African Americans say that religion is an important part of their lives compared with less than half of whites and only slightly more for Latinos. Black millennials are far more religious than their white and Latino counterparts. Over 60 percent say that they pray daily and almost 40 percent attend church service weekly. They tie a sense of peace and wellbeing directly to spiritual and religious engagement.

While not as high as their parents and older African Americans, the point is still important. Those of millennial and Generation X are more likely to believe and act on climate and environmental issues as opposed to older Americans. They are more likely to believe the science around climate and find ways of linking it to their own religious beliefs instead of separating them altogether.

We go to church weekly, pray regularly and believe with an undeniability that God is real. I don't know how many times I've heard the phrase, "You might not believe in God but that doesn't mean God doesn't believe in you," repeated at the end of church service during the call to salvation for sinners seeking redemption. When I was a student at Spelman College, it was hard for a few of my women and religious classmates to embrace different ideals about God, Jesus and biblical teachings. The idea that God "might" be a development of man's way to personify spirituality was met with harsh criticism from students who'd spent their entire lives believing the opposite. How can you tell a room full of the nation's top educated Black female students including pastors' daughters, that God might be a figment of their imaginations and not invoke major side-eyes? One student walked out of the class vowing to pray for the professor as she went to call her father and tell him of the blasphemy.

But when you think about it, it makes sense. The African American community has suffered atrocities that required focus and meditation on a higher power to survive. The scripture is important because it tells a story familiar to our own. It's a story of slavery and hope for redemption. It describes salvation through the process of being adopted and grafted into the family of God. It's the literal description of being cast aside but then later determined to be survivors—a chosen and special people. It's an ongoing desire to be wanted by someone—anyone, and who better than to be chosen by than the King? How wonderful a story to be considered a slave that travails through torture unknown only to become ruler of them all? These are the stories of hope and redemption that our ancestors clung to and has sustained us despite continuing to suffer from the injustices of a country that has not yet accepted us as equal. It is the hope of green pastures and still waters while living alongside landfill dumps and water contaminated with lead. To experience White Christian America, a group that proselytizes as a faith that speaks of love and freedom but practiced hate, violence and bondage against Black Americans—the fact that Black Christians have remained Christians is in and of itself a miracle. From enslavement to inequity, religious connection to community has been seen as both a constant comfort for some and a launchpad to justice initiatives that Jesus himself outlines as our one true commandment: to love one another.

Today, Black church leaders are embracing this close connection and identity between climate, the environment and biblical teaching. The spread of Green Theology—the teaching of biblical tenants of faith as the Christian's responsibility to protect and preserve the environment, is becoming widely accepted in the Black church. Rev. Mariama White-Hammond, the founding pastor of New Roots AME Church in Massachusetts, is a powerhouse faith leader known for her ministry work interweaving ecology, justice and faith. In 2021, she was appointed Chief of Energy, Environment and Open Spaces for the City of Boston. Imagine that, a Black woman pastor in charge of energy and environment for the entire city of Boston. Organizations such as the National Black Church Initiative, led by Rev. Anthony Evans, The People's Council led by Rev. Michael Malcom and Green the Church founded by Rev. Dr. Ambrose Carroll, are working to turn scriptural anecdote into environmental action while connecting the dots between social justice and climate action in the church community.

I met Rev. Dr. Ambrose Carroll at the Congressional Black Caucus in 2017. We were both speaking on a panel on the Black church and the environment and as the incoming National Field Director for Moms Clean Air Force, I was sharing about our upcoming bible study entitled "Breath of Life." I will never forget Dr. Carroll's passion and conviction about why the Black church had to be the centerpiece of discussing climate issues in the Black community. He spoke of the Black church as a "sleeping giant" that once awakened, could invoke mountain-moving power with respect to climate policy and equity in the Black community. Though I'd read about Rev. Carroll many times, I was enthralled after hearing him. He got it. For generations, the Black church has been the healer of multiple traumas experienced by the Black community. The Black church has been the centerpiece for social justice movements in the African American community. It would make sense that the Black church would and should be the conveyor of effectual change in the Black community when it comes to climate change and environmental justice. The Green the Church initiative is designed to be just that—a place to host resources, amplify a green theology, promote sustainable practices and build power for political and economic change. Green theology emphasizes the Christian's duty to protect God's creation in the same way that we needed someone to protect us. Creation care ties our own prosperity to that of the earth's and helps parishioners understand that if we take care of what God has blessed us with, he will, in turn, take care of us. The activity of creation care is exhibited by promoting sustainable practices such as responsibly using energy and identifying renewable energy as a respectful and wise way of using God's gifts of sun, wind and water to restore our communities. Building the health and well-being of the earth in turn builds the health and well-being of the church. Sustaining these practices must be tied to building the political power that influences local, state and federal policy. Faith can support governmental transformation to a greener economy while creating more resilient communities.

BOSOM NOTES

"Listen baby, stick this in your bra, I want to make sure you don't lose it."

Only items of precious value and importance are kept in a Black woman's bra. Be it a hidden $20 bill, the phone number of a special person or numbers to be played in the lottery, there is no more secure place on this planet. To be given a note to put in one's bra, close to the heart, is an unspoken message of trust, and this dear lady was communicating it clearly. My colleague and I exchanged knowing glances and smiled as the older African American lady, clad in her Sunday's best, handed her a folded-up piece of white paper. On it was written her name, address and phone number. It was 2018 and I was presenting while she tabled at the National Women's Auxiliary of the world's largest African American Baptist Organization—the National Baptist Convention. The National Women's Auxiliary wanted to get armed and ready to fight climate change in the name of Jesus. We were sharing the Moms Clean Air Force climate-focused Bible study, but we'd made a huge miscalculation. In a room of over 350 members of the Women's Auxiliary leadership, we ran out of Bible study books. This woman wasn't leaving without the assurance of getting one. With a stern yet graceful expression, she patiently stood by and watched as my friend replied to her with a simple, "Yes ma'am," and stuck the paper in her bosom. In less than five seconds, a generational transaction had taken place, and the assurance of its fulfillment carried the weight of sustaining the future of our people.

Black folks, especially Black women, have always had a physical and spiritual connection to the environment. Our history has entwined us with it in a profound way and our connection to the land is as symbiotic to us as bees to flowers. Time and time again, our voices are constantly ignored on matters concerning climate impacts and environmental protections. That's why meeting with the women's auxiliary group was exciting; not only was this the National Baptist Convention opening doors for a discussion on environment and spirituality, it was Black women seeking to advance their voices on climate. We know what a real emergency looks like. I stood in awe watching these Baptist church mothers, joined by our faith in God, focused on increasing the awareness of climate issues in our communities. For the first time, she was hearing about things she could do in her faith and her community to lessen the impacts of climate change. She knew that this was knowledge in a familiar voice, yet one missing from network news and print media. She knew that if we didn't use this precious moment to stay connected, even if by a small piece of paper, we may lose the chance to share it with others. She stood by with an unwavering persistence to make sure we got her the materials she needed.

It's what makes the words and work of Rev. Carroll and others easily accepted as a familiar reminder that Black people of faith have a special role to play in solving the climate crisis. It's because time and time again we've applied our belief in what scripture calls mustard seed faith and we've seen the manifestation of our seed turn to fruit. Jesus makes reference to the mustard seed twice in the New Testament, but it's the story in Matthew that relates closely to the relevance and need for solutions from Black churches as part of the climate crisis today.

In the book of Matthew, there's a story about the inability of the disciples of Jesus to heal a child due to their lack of faith. Jesus's disciples were doing their best to follow the example of the teacher but came upon a situation they couldn't handle. To make matters worse, Jesus had to hear about this failure from the boy's family instead of his own followers. I imagine Jesus had to be a bit perturbed when the desperate father came to him and said, "Jesus! We need to speak to management. We tried going first to the folks you've got helping you and, bless their heart, they obviously need a refresher course in healing." I imagine Jesus gave his disciples an epic side-eye. According to Scripture, Jesus said to his disciples, "You unbelieving and perverse generation, how long shall I stay with you? How long shall I put up with you? Bring the boy here to me." Jesus healed the child. But the disciples pulled the Lord to the side and said, "How come we couldn't do that?" In verse 20, he replied, "Because you have so little faith. Truly I tell you, if you have faith as small as a mustard seed, you can say to this mountain, 'Move from here to there,' and it will move. Nothing will be impossible for you" (NIV).

A mustard seed is one of the smallest seeds and grows fast and strong. Among Christians, the mustard seed is not only a symbol of faith, but also of growth, tenacity and determination. For Black Christians, it's a symbol of hope and legacy. To be the smallest of the small, the lowest of the low, buried in the ground, broken, bruised and torn apart is symbolic of both the experience of the seed and Black America. But, if like the seed, we trust

God in the process of growth, development and expansion, we too will see our children and our children's children arise strong and resilient. That is the spirit of mountain moving, demon dispersing, climate shifting, mustard seed faith. It is the lived experience of Black Christian Americans every day. As we watch the climate crisis worsen across the globe and mountains that are moving, the mainstream environmental movement needs a bit of faith and hope as part of the solution. If there's any group on earth that can be assured and trusted to serve in that part of the solution, it's praying Black folks with mustard seed faith. That is what our ancestors are counting on.

MAKE IT MAKE SENSE—CLIMATE AND FAITH TERMINOLOGY

Creation Care

Creation care is the responsibility of humanity to care for and be responsible for the growth and development of all of God's creations including the people, plants, animals and all-natural resources on Earth. Creation care recognizes God's charge to Adam and Eve in the book of Genesis as a call to stewardship and accountability for the multiplication of God's original creation.

Green Theology

Green Theology views the Bible through the lens of God's relationship with all of creation including nature, the environment and humanity. It examines the responsibility of Christians to consider the planet itself as a spiritual part of our existence and a charge to provide it the same justice and freedoms that we seek through our faith. In Green Theology, the outcome of humanity is intricately connected to the outcome of the planet physically and spiritually.

BEFORE THE STREETLIGHTS COME ON

1. It's okay to love Jesus, Buddha and the Prophet Muhammad—and science! Read the "Breath of Life" Bible Study by Moms Clean Air Force.

2. Part of a faith tradition? Join a faith-based climate action group. Green the Church and Interfaith Power & Light are two dynamic groups committed to creation care among the diaspora.

3. Listen to the naturalist in your family. These are the men and women we've often termed as witches and hoodoo folks. Take a moment to respect the stories. Many come from places of love and the shared experience of paying homage to nature as our host and our guide.

4. Teach a scripture series, create a song, write a play, do anything creative that tells the story of connectivity to nature and the environment through faith.

5. Tithes, offering, charitable donation—GIVE to faith-based climate groups.

6. Vote for people and policies that care for creation instead of tearing it down.

CHAPTER 12

A SOUL-POWERED FUTURE—THE FUTURE OF CLIMATE OPPORTUNITIES THROUGH CLEAN ENERGY, CLIMATE CHANGE AND THE NEXT GENERATION OF SOLUTIONS

Hear me out on this one.

What do Beyoncé, South African rap artist Da L.E.S and my seventy-plus-year-old dad have in common (besides the fact that they're all Black)?

They all drive electric vehicles, Teslas to be specific. That's right, the Queen Bey herself is "Flawless" in her Model X with Jay-Z and the kids. Da. L.E.S dropped a single entitled, "Elon Musk" complete with video and the car highlighted. Meanwhile, my dear, sweet, civil rights fighting, teacher trained, retired parents are happily running errands in their 2021 Tesla Model X Long Range. It was with immense joy that my dad expressed how driving up to the charging station, he saw other young Black people, playing hip-hop, first politely putting on their masks before approaching and asking about his car. My absolute favorite TikTok video is of @championxiii, a young artist of color with tattoos and kinky twist, riding down the highway with a blanket and what looks like Waffle House eggs and bacon, in a Tesla on autopilot. If that is not the epitome of Generation Z, I don't know what is. While this is not an endorsement of Tesla, it is an example of the trend toward electric vehicles. The Ford F-150, the most popular truck in America, now has an all-electric version that can pull a train and power a house. Regina King (our favorite teen from *227* turned badass police protector in *Watchmen*) replaced Matthew McConaughey as the commercial lead for the electric version of the Cadillac. There has been a seismic cultural shift in the acceptance of alternative fuel and climate-friendly vehicles in the minority community and it spans both generations and geographies.

It also spans jobs, a green economy and a chance to experience a wealth transfer like never before. Likewise, The EV movement could benefit from a little "soul" power, Black people set trends, direct habits and cultivate solutions alongside fellow minority groups that leverage the intersectionality of social issues with profit. If we take full advantage of this next industrial-style revolution—clean energy investment, climate innovation and adaption, alternative fuels—and begin to advance environmental justice and infrastructure stability in Black and brown neighborhoods, we can transform a generation: one charging station and electric vehicle at a time.

The African American buying power in America continues to increase and we are anything if not loyal to a brand. At the same time, social justice issues like climate change and racial equity matter to us and we will spend our dollars accordingly. The 2020 Neilson report showed that despite COVID-19, African Americans did 11 percent more shopping online than the average household, are three times as likely to show support of their favorite brands on a social media platform, and expect those brands to take a stance on issues including the planet, social justice and wellness. To top it off, it is estimated that between 2020 and 2060, Black Americans will add more than 20 percent to the total US population growth. In other words, Black folks are carrying a whole lot of untapped leverage and the environmental industry would be wise to take note.

We set trends and we can make or break it at a moment's notice. Ask any Black household about that purple velvet bag with the gold rope and I guarantee recognition because somebody's uncle, dad or cousin collected the "Crown Royal bag." The royal blue tin that was supposed to hold four variations of vanilla sugar cookies shaped like a circle or pretzel? That was a barrette holder or sewing tin. Or how in the late 90s the massively destructive rumors that spread about Tommy Hilfiger and that he didn't make clothes for Black people? I vividly recall NEEDING the red, white and blue oversized sweatshirt with Hilfiger stretching from one shoulder to the other. Everybody from TLC to Snoop Dogg on *Yo! MTV Raps* was wearing Tommy. All of a sudden, we heard rumors within the Black community that he was racist. People started saying that he went on *Oprah* (of all places) and she threw him off the set after he made a claim that he didn't make his clothes for Black people. Although it was a total and complete lie, Black people believed it. Anything related to *Oprah* was tantamount to gospel and in 2007 she hosted him on the show to set the record straight. The damage had been done and his brand hold on the hip-hop community and their audience was gone. Today, you can buy Tommy Hilfiger at Ross Dress for Less.

Now let us apply that same logic to vehicles. At age eleven, I was bopping around the house with a pretend towel-wig wrapped around my head and a hair comb for a microphone singing along to Pebbles's "Mercedes Boy" as if I had my own Benz and the checking account to match. In 1985, Lord knows I was not supposed to be listening to Prince's *1999* album, but I was nine, and honestly, who didn't know the lyrics to "Little Red Corvette"? Along with Aretha Franklin's "Freeway of Love," my dad and I would alternate between the choruses of "Pink Cadillac" and "Little Red

Corvette" while riding down the street. Let's not forget that the way to spot a solid Mary Kay consultant (or run in the other direction) was the pink Cadillac slowly rolling into a neighborhood, products in tow. As I got older, the connection between vehicles and success continued. Countless hip-hop songs included specific definitions of luxury cars. In 2003, the radio stations played Lil' Kim's "The Jump Off" in heavy rotation with a solid chorus:

"This is for my peeps, with the Bentleys, the Hummers, the Benz, Escalades twenty-three inch rims, Jumpin out the Jaguar with the Tims."

It seemed like every hip-hop video for the next ten years featured one of these cars, regardless of affordability or the gas-guzzling expenses associated. But by 2011, when electric vehicles began gaining attention, there was no real marketing to minority communities. The Chevy Prius, GM Volt, and Nissan Leaf were geared towards middle-income white people based on the idea of affordability. A happy white couple heading out for a hike in their Volt. The young white college graduate carpooling in the HOV lane in his Prius. The images felt like strange nods to what the future should look like concerning renewable energy, but through an eerily familiar lens void of color and diversity. How did you fuel or energize these things? Charging stations were scarce and certainly not in the Black community. Did the research and marketing teams not look at the data of what Black people were driving? Did they hold any diversity in the first place? No offense but no one wants to hear, "Hey man, wanna go for a ride in my Leaf?"—not sexy.

Now the tables have turned. Studies have shown that Black Americans and Latinos are more concerned about climate impacts than any other demographic. Not only are our communities concerned, we are working to intersect the complex web of social justice dynamics with climate change by electing leaders that support strong and equitable climate policy, investing in sustainability and incorporating celebrity influence. I was floored when someone sent me a message that famous Black actress, Regina Hall, had listed me as her #WCW (Women Crush Wednesday) on Instagram.

"Girl! Do you know Regina Hall just posted you on the Gram?" my best friend exclaimed. "I didn't know she was an environmentalist too!" We both chuckled at the fact that climate change impacts on the Black community are not lost on Black celebrities and entertainers; they came from the same highly polluted environmentally impacted neighborhoods as the rest of us. Some celebrities are just now connecting the dots because they too are seeing themselves in with the activists, scientists and CEOs solving the problem. Global entertainment artists Beyoncé and Megan Thee Stallion are both from Fifth Ward Houston, Texas, one of the most polluted communities in the state. Rapper Lil Boosie is from Louisiana's Cancer Alley. Actor Don Cheadle is not only a United Nations Goodwill Ambassador advocating for the reduction of single-use plastics, he started his own organization along with noted actors George Clooney, Matt Damon and Brad Pitt. This is not light work and the influence of Black Americans can accelerate the acceptance and implementation of climate strategies. Combined with the increase of Black buying power and the emphasis from the Biden administration on clean energy, environmental justice and infrastructure development, Black communities have a clear shot at leveraging our support and thereby strengthening our communities economically, socially and culturally, all in the name of climate action.

The work has already begun. The beginning of the Biden-Harris presidential administration saw the creation of the WHEJAC—The White House Environmental Justice Advisory Council. A gathering of thirty of the nation's most established and profound leaders on environmental justice, the group was tasked with the mission of advising the White House on what it will take to establish equity climate policy and see it implemented throughout the country while addressing past environmental injustices. Leaders on the council include noted Black American environmental justice practitioners like Dr. Beverly Wright of Louisiana, Dr. Robert Bullard of Texas, Catherine Flowers of Alabama and Harold Mitchell of South Carolina. Their efforts helped push the executive orders signed by

President Biden in 2021, ensuring that federal infrastructure investments are based in clean energy instead of fossil fuels with a focus on addressing environmental disparities that have existed in disadvantaged, often Black and brown, communities. In addition, the Biden administration created the Justice40 program, an initiative to push 40 percent of the climate and environment investments to the most vulnerable and impacted communities first. Employing an "all of government approach," every agency of the federal government is tasked to incorporate climate actions into the daily mission and push funding to the front line, marginalized communities that need it most. The rapid impacts of extreme weather are proof that we need these actions quickly.

The prospects around the use of electricity as both a clean fuel alternative for the power grid and automobile sector are a necessity after the Texas winter storm of 2021. Who among us will forget the images of Texas citizens freezing while their leaders escaped to sun and sand? There were downed powerlines and no electricity. City government buildings and schools shut down because of impassable roads. Waterlines froze and water mains broke due to the weight pressing down on already overburdened and outdated infrastructure systems. Forced burn offs from emitting facilities put tons of dangerous pollutants in the air. But Senator Ted Cruz, one of the biggest deniers of the climate crisis, was caught on video escaping to Mexico. He left the family dog, Snowflake, in the cold house. Meanwhile, the fossil fuel industry and its supporters in the state doubled down. In an effort to shift blame to renewables, *Fox News* and *Tucker Carlson* promoted inaccurate pictures of frozen windmills in Texas as the cause of outages (they were actually pictures from Sweden). The smoke and mirrors of it all is no accident: fossil fuel industry giants know full well that this colossal failure of the Texas energy grid is indicative of a much bigger problem. It put a huge spotlight on the fight around renewable energy and electrification infrastructure and it will be people of color leading that fight. While the fossil fuel industry has argued that electrification is unreliable and a disaster waiting to happen, it is particularly hard to defend when homes went without power for days.

There's no question that communities of color and marginalized people are bearing the brunt of this failure. The "rolling blackouts" were never rolling; in Black and brown communities they sat and stayed. My friends started posting pictures of the text messages they received from Entergy, asking them to reduce their electric usage to protect the grid, and it was like hearing a bad joke. "Are they gonna reduce my bill?" was a common response as people struggled to figure out how they were going to stay warm and eat, let alone decide what appliances were "nonessential." The idea of making that decision comes from a place of privilege. It's not like folks are choosing whether to unplug the Instant Pot or the coffee maker; when you only have the basics, every appliance is essential. And their fears were real: families received bills 20 times higher than normal. Can you imagine having a regular bill of $150, then seeing that your provider has sent you a bill for over $3,000 while you have no power?

All of the social, environmental, and infrastructure problems exposed in this winter storm make clear that energy inequity is real and must be addressed in a way that fills the holes created by systemic racism and lack of resources. Climate change impacts extreme weather, and it can show up in the form of hurricane, wildfire, fire, and yes, snow and ice. The climate crisis may attack the same geographic area but the way the people of different demographics and socioeconomic backgrounds experience that attack is widely different. It highlights the reason we must engage, as a community of people, to plan, prepare and participate in efforts that push for a strong infrastructure plan that directs dollars into the most impacted communities first and fast. We must leverage local community power alongside business and philanthropic willingness to effect systemic change.

A $2 trillion investment proposed by the Biden administration, the largest of any administration, to cities and towns across the country would create jobs and economic opportunities and provide long-overdue access to economic opportunities by way of acting on climate. The leverage opportunity from philanthropy is even greater. The Bezos Earth Fund, Bloomberg, The Walton Foundation, are among others that have committed their charitable dollars to climate but have a mandate to include addressing the issues of environment and climate justice. The inclusion of diverse experts and efficiency of creative ideas must happen to show proof that intersecting justice and climate can work.

Back home in the neighborhood, it means that the abandoned basketball court next to the old gas station in town can be revitalized into a greenspace, with courts made from recycled materials, while acting as a hub for internet service that's connected to an electric school bus that powers the lights to the basketball court while it is idle in the middle of the day. It's the chance to normalize charging stations as an infrastructure standard, seen at both Whole Foods and McDonald's. It is a chance to offer subsidies to first-time car buyers that move them towards EV cars versus gas and economic dependence as opposed to the idea that electric public buses in Black neighborhoods are sufficient. It's getting the front parking spaces at church on Sunday because that's where the EV charging stations are located. It's watching HBCUs engage in green investment for students and sustainable building practices to anchor climate adaptation in the community. This is a pivotal moment to honor the history of how Black Americans encounter the environment, challenges and collective culture while embracing a Black future filled with opportunities through alternative fuels and clean energy.

In the Bible's book of Esther, the story of a Jewish girl turned queen reveals how one's purpose can be designed to shift humanity. Formerly known as Hadassah, the Jewish girl was forced to set aside her name, her place and her people to adapt to a new and different environment, all at the will and pleasure of a king. Hadassah, turned Esther, flourished but not without controversy and hesitancy of engagement on issues that concerned the wellbeing and future of her people. In a moment of frustration, her uncle had to have what I'd like to call a "Come-to-Jesus" meeting with her. In Esther 4:14 (ERV), he states:

> If you keep quiet now, help and freedom for the Jews will come from another place. But you and your father's family will all die. And who knows, maybe you have been chosen to be the queen for such a time as this.

Either Esther could speak up, participate and advocate on behalf of the Jewish people or she could risk death along with everyone else for she knew that she would not be exempt because of her status. I feel that Black Americans are in our Esther moment. Despite the trauma of our history, struggle, separation, abuse and neglect, we have purpose and power for the world that has yet to be fully embraced. Who knows? It could be for such a time as this. What we do know is that without our engagement to secure justice, participation with ideas for innovation and creativity for influence, the survival of all people is at risk. A little "soul" may be just what the entire world needs in order to see hope and resiliency for a strong climate future ahead.

MAKE IT MAKE SENSE—ENERGY TERMINOLOGY

EV (Electric Vehicle)

Cars and trucks powered solely by electricity instead of gas. The biggest difference is how they generate power. Gas-powered cars have moving parts and are powered from the heat igniting the engine. EV cars and trucks are powered by battery similar to the way your cell phone works. Charge it up and it runs until you are out of battery power. The downside of the EV is that it takes a while to charge it up again whereas gas powered-cars are a quick stop in the gas station. On the other hand, the fuel economy of EVs is more cost-efficient. Think of it this way, to go a full 385 miles would you rather fill up a fifteen-gallon tank at $4.00 a gallon or spend a hefty $19.00 to charge your car?

There are three basic types of EVs. Battery electric vehicles or BEVs run on battery power with no gas at all. Examples of BEVs are Tesla and the Ford F-150 Lightning. Plug-in hybrid electric vehicles (PHEV) use a mix of gas and electricity. Typically, the car starts out on electricity but can shift to gas if needed. Toyota Prius and Jeep Grand Cherokee 4XE are examples of PHEV. Finally, the Hybrid Electric Vehicle (HEV) runs mostly on gas but use electricity to reuse power when breaking. This is called regenerative braking, and you feel it every time you drive and the car seems to turn off when you stop. Toyota Camry and Honda Civic have cars that are HEVs.

Energy Grid

An energy grid, also known as an electricity grid, is the system that gets power from the source to your home or business. Power is generated from multiple sources including renewable sources like solar panels or windmills. There is also the burning of fossil fuels and creation of energy from nuclear reactors. The site of energy creation is known as the generator or power plant site. After it's created, the power must travel to a local point for distribution. The travel process is called transmission and it crosses our country overground through poles and wires and underground through buried lines. Once the energy arrives at the local distribution point, it is converted as alternating currents (AC) to be delivered to the outlets in our homes. The grid is like the movie *The Matrix*, everyone is connected and it's all around you, unless you intentionally choose to be off the grid.

BEFORE THE STREETLIGHTS COME ON

1. Consider a new electric vehicle or used gas-powered vehicle for your next car. Do NOT purchase a new gas-powered vehicle. The future is EV and cars are being made faster and more efficient and can do multiple things. The F-150 Lightning truck can power your house. I'm here for it!

2. Consider businesses as expansions of the EV transportation network. If there are going to be thousands upon thousands of EV charging stations, somebody has to fix them right? Businesses that involve the setup and maintenance of electric charging stations are a great way to start.

3. Do you have land? Access to land? Family land that no one knows what to do with yet? Think about turning it into a solar farm. Solar is a renewable energy source that can be stored in batteries and used at a later time. Solar farms are an excellent way to repurpose land like brownfields or sites that may not be fit for use of another purpose. Create power for yourself and others.

4. Let's power the streetlight with renewable energy!

5. Vote for people and policies that provide incentives for buying electric and subsidies for those who may not yet be able to afford them. EVs shouldn't be for the privileged alone. Subsidies can help families obtain an electric vehicle and take advantage of a whole host of opportunities that were previously unavailable. A child without Wi-Fi in the home can connect online in an electric car with Wi-Fi.

ACKNOWLEDGMENTS

First giving honor to God, who is the head of my life...

Some of you read this line and chuckled a bit, completing the sentence in your own way. If you know, you know.

But I must start with a sincere heart of gratitude to God, the Divine, the Holy Spirit, all the angels and ancestors that encouraged and protected me along this writing journey. This book is a physical manifestation of faith and evidence of hope realized.

Faith also took the shape of a fine, strong, supportive husband who has both the patience of Job accompanied by the humor of Bernie Mac. Without the constant assurances that ranged from "Baby, it's okay" to "Get yo azz on and go write," this book would have sat dormant in my brain for another ten years. Between work and late-night writing, managing a three- then four- then five-year-old and a teenager during COVID-19, this man deserves sainthood. Dexter L. Toney, thank you. I love you and there's no one else I'd imagine riding this wild ride called "life" with other than you. It's me and you against the world boo!

Thank you to my beautiful children, Devin, Deriah and Dexter. Devin, your smile and joy brought so much light to this work. I knew I could depend on you for a laugh every day without fail. Deriah, so much of you is in each of these pages. Thank you for your witty insight and creative way of thinking. I love you more than you know!

To my parents, Mercidees and Charles Victor McTeer, and my brother, Marcus McTeer, thank you for encouraging me to live well and walk in the knowledge that justice and freedom are something we must not only fight for, but work to keep. Seeing your work and commitment to the Mississippi Delta taught me that being a Black American is a gift and we are special. Thank you for always being the future we should strive to become.

Asia Guest and Lori Wilson, you both stepped in and helped out in ways I will forever be grateful for. Our family is grateful for the care you showed us always. Thank you!

My DOT friends, Von, Shayla, Lilly, Halima, Crystal, Lisa and Shaunna, it goes without saying that your check-ins and our getaways kept me sane. Thank you!

Devi Lockwood, Renee Ferguson, Dominique Browning, Heather Woods Rudolph, Dr. Katharine Hayhoe and my *DAME* magazine family, thank you for being honest and true readers and editors for my work. Many thanks to the members of Rachel's Network. You asked how you could be helpful in fulfilling my dream of writing a book then followed through with steady assistance. Your encouragement to write and keep writing is the reason this book exists.

My editors, Lil Copan, Adrienne Ingrum and Rachel Reyes, thank you for putting up with the questions, adds, edits, deletions and antics of a novice writer. Your expert guidance and genuine feedback are exactly what I needed. Thank you and the entire Broadleaf Books and 1517 Media family for pushing me to greatness!

Tremendous thanks to the City of Greenville, Mississippi, the US EPA, Moms Clean Air Force and the Environmental Defense Fund. Working with such amazing people and leaders in the environmental movement has had a profound impact on my life. Each organization is proof that working together can really work.

My best friend of more than 25 years Jackie Thompson, Margot Brown, Dr. Robert Bullard, Peggy Shepard, Dr. Beverly Wright, Sharon and Shamyra Lavigne, Mustafa Santiago Ali, Lisa P. Jackson, Karan Kendrick and Mr. Jimmy Smith of North Birmingham—thank you for sharing your stories and helping me tell mine. To all my church family and friends, sorors of Alpha Kappa Alpha Sorority, Inc., cousins, aunts, uncles, grandparents, in-laws, classmates and friends, thank you for the role you've played in my life and the future of our collective communities.

And thank YOU for taking the time to read my thanks and this book. Let's go together and create solutions in gratitude.

Heather McTeer Toney

STREETLIGHT ACTION PLAN (S.L.A.P.)

Educating ourselves and each other about climate impacts on various communities empowers us all to find solutions that work. Now continue the work and create a plan to take action before it's too late.

Here is a list of 50 things you can do today to talk about climate action that centers on equity and experience for culturally competent climate solutions. Now go and share it with friends before the streetlights come on.

1. Talk about it. Take any and every social justice issue that concerns you, insert climate change and it may or may not impact your concerns.

2. Normalize climate conversations every day in every way.

3. Share the stories from your family that connect you to climate, nature, the environment and survival.

4. Read the NAACP's *Fossil Fueled Foolery* reports 1 and 2.0.

5. Identify your public service commissioner or energy cooperative association and representative.

6. Identify any environmental impacters in your community. This may include oil refineries, petrochemical facilities, natural gas industry, CAFOs (Concentrated Animal Feeding Operations) or a landfill. If you are in the vicinity of a methane emitting facility, identify and ask for the plant plan to reduce methane emissions.

7. Make sure there are renewable energy and climate aspects of the science fair at your local school.

8. Look at the list of redlined city maps that can be found on the Digital Scholarship Lab website to determine if your city was a redlined

community. Ask your local and state leaders what (if any) changes have been made to ensure compliance with federal fair housing practices.

9. If you live in an area or subdivision subject to housing covenants or homeowners association, check the governing document of the organization and group for accommodations to ensure equitable and future sustainable reliance on renewable resources.
10. Identify and know members of your local housing authority and planning commission. Consider serving.
11. Prepare playgrounds with green spaces in mind and plan healthy places to play.
12. Implement resiliency and adaption plans for climate impacts at public housing development projects.
13. Read the National Equitable and Just Climate Platform and become a member.
14. Attend and stay up to date on local boards and commissions that have oversight, influence or provide resources to community planning and development. These include:
    - City planning and zoning commission
    - Insurance commissions
    - Community Land Trust boards
    - Realty/Realtor boards
    - Chamber of Commerce
    - City and County Redevelopment Commissions
15. Share the story of your neighborhood, family or community presence in the area and how it has changed due to climate and environmental impacts.
16. Identify a school and get involved with the long-term resiliency planning for future development in the community.

17. Twitter, TikTok, Instagram. Use all the social media tools and add climate hashtags.

18. Plant, grow and farm.

19. Utilize regenerative and restorative farming practices that are beneficial for people and the planet.

20. Look for ways your community may be unknowingly paying the EPT—Environmental Poverty Tax—on food and water.

21. Invest in and advocate for cleaning water sources instead of constantly donating water to communities with unhealthy water supplies.

22. Eat less meat. Find creative alternatives to meat and meat byproducts.

23. Patronize black plant-based businesses, influencers and creators.

24. Support HBCUs, specifically agricultural and mechanical schools with a focus on food production and equity.

25. Donate to HBCUs.

26. Identify all potential sources of toxic chemical releases within your zip code. The TRI Map can be found on the EPA website at www.epa.gov in a section entitled, "Where You Live."

27. Items in your home and car emergency weather kit should serve for environmental protection. Masks, gloves and safety glasses. Include battery backup, adapters, protective clothing and hats.

28. Encourage young people and youth-affiliated organizations to develop pollution prevention and extreme weather resources for the community.

29. Invest in a home generator and look at alternative energy sources as backup power for your home or business.

30. Participate in your local school board and ask for the long-term planning for climate adaptation and building resiliency.

31. Look for energy efficiency opportunities in your school or educational institution.
32. Create a student-run solar farm.
33. Explore outdoor classrooms and opportunities to get students outside and into the elements of air and nature if possible.
34. Create a school garden or create a class and teach students farm-to-table practices.
35. Design natural safe spaces that include air, water and land.
36. Plant trees indigenous to the area along the sidewalks and driveways to the school building.
37. Join your local community policing initiative. If you don't have one, create one.
38. Plan for summer and extreme heat resources to remind the community of the dangers associated with extreme heat. Plan ahead by providing summer and/or heat relief resources for the community. Cool off, open pools and creatively reuse water with splash pads.
39. Talk to your local police department or sheriff's department about lighter clothing for officers in the summertime.
40. Encourage law enforcement on bikes and walking as opposed to idling engines.
41. Support alternative dispute resolution sessions in both law enforcement and community action centers and schools.
42. Arrange activities on grass instead of asphalt.
43. If you have land or access to land, repurpose with renewable energy investments. Create power for yourself and others.
44. Part of a faith community? Join a faith-based climate action group. Green the Church and Interfaith Power & Light are two dynamic groups committed to creation care among the diaspora.

45. Tithes, offering, charitable donation—GIVE to faith-based climate groups.

46. Listen to the naturalist, non-religious people in your family. They have ideas about protection and shared space with our environment.

47. Advocate for EV charging stations in your community. Put one at your place of business, place of worship, school or community center.

48. Consider a new electric vehicle or used gas-powered vehicle for your next car.

49. Advocate for zero emission vehicles (ZEV) like school buses, city transit and rail in your community.

50. Vote!

    - Vote for people and policies that provide incentives for buying electric and subsidies for those who may not yet be able to afford them.
    - Vote for sustainable restorative climate policy.
    - Vote for people who support strong climate supportive policy.
    - Vote for people and policies that reduce methane and greenhouse gas emissions and center equity as part of the solution.
    - Vote for local and state goals with targets of reducing greenhouse gas emissions by 2030 or 2040.
    - Vote for people and policies that include local community members and culture as a matter of practice.
    - Vote for people and policies that support equity in food production and distribution.
    - Vote for people and policies that operationalize climate adaptation through infrastructure and jobs.
    - Vote for school board members and policies that are dedicated to educating students in safe and resilient school buildings.

- Vote for community-led, community-driven policing initiatives.
- Vote for proactive learning exchanges that are more effective and safer than defensive responses to violence.
- Vote for people and policies that care for creation instead of tearing it down.
- Vote for voting. Whatever you do, vote. VOTE.

NOTES

CHAPTER 1

5 ***"Black people make up 13 percent"***: "Disparities in the Impact of Air Population," American Lung Association, https://www.lung.org/clean-air/outdoors/who-is-at-risk/disparities.

5 ***"We live in areas four times as likely"***: "Disparities in the Impact of Air Population."

9 ***"Inflammation of the lungs"***: Professor Stuart Anderson at the London School for Hygiene and Tropical Medicine

10 ***"In a study entitled"***: "Chronic Obstructive Pulmonary Disease in America's Black Population," *Pulmonary Perspective*, https://www.atsjournals.org/doi/pdf/10.1164/rccm.201810–1909PP.

10 ***"Environmental factors and climate fluctuations"***: "Climate, Ecosystem Resilience, and the Slave Trade," VoxEU, https://voxeu.org/article/climate-ecosystem-resilience-and-slave-trade.

11 ***"Due to where enslaved Africans were brought"***: "Majority of African Americans Live in 10 States; New York City and Chicago Are Cities with Largest Black Populations," United States Census Bureau, https://www.census.gov/newsroom/releases/archives/census_2000/cb01cn176.html.

11 ***"As temperatures rise"***: "Climate Risk and Spread of Vector-Borne Diseases," Climate Nexus, https://climatenexus.org/climate-issues/health/climate-change-and-vector-borne-diseases.

11 ***"By way of agriculture"***: "Rice Reveals Enslaved Africans' Agricultural Heritage," SAPIENS, https://www.sapiens.org/culture/african-rice-new-world.

11 ***"The 1890 land-grant institutions programs"***: "1890 Land-Grant Institutions Programs," National Institute of Food and Agriculture, https://www.nifa.usda.gov/grants/about-programs/program-operational-areas/1890-land-grant-institutions-programs.

12 ***"In 1982, students protested"***: "The Environmental Justice Movement," NRDC, https://www.nrdc.org/stories/environmental-justice-movement.

12 ***"After helping Dr. Martin Luther King Jr."***: "SNCC," HISTORY, https://www.history.com/topics/black-history/sncc.

12 ***"After helping Dr. Martin Luther King Jr."***: "Who Was Ella Baker?," Ella Baker Center for Human Rights, https://ellabakercenter.org/who-was-ella-baker.

12 ***"The housekeepers and janitors in South Louisiana"***: "Climate Change Is Also a Racial Justice Problem," *The Washington Post*, https://www.washingtonpost.com/climate-solutions/2020/06/29/climate-change-racism.

12 ***"The housekeepers and janitors in South Louisiana"***: "What Is Cancer Alley?," Verywell Health, https://www.verywellhealth.com/cancer-alley-5097197.

12 ***"Why did it take watching the police kill George Floyd"***: "Justice in Every Breath," Moms Clean Air Force, https://www.momscleanairforce.org/justice-in-every-breath.

13 ***"These experts"***: Dr. Robert Bullard, from Texas Southern University, is commonly referred to as the Father of Environmental Justice. Dr. Mildred McClain of Savannah, Georgia is a noted environmental justice advocate who has served the coastal community in Georgia for over forty years. Dr. Beverly Wright is the co-founder of the Deep South Center for Environmental Justice.

13 ***"2020 was one of the hottest years"***: "NASA Says 2020 Tied for Hottest Year on Record," *Scientific American*, https://www.scientificamerican.com/article/2020-will-rival-2016-for-hottest-year-on-record.

13 ***"2020 was one of the hottest years"***: "COVID-19 and Climate Change Threats Compound in Minority Communities," *Scientific American*, https://www.scientificamerican.com/article/covid-19-and-climate-change-threats-compound-in-minority-communities.

13 ***"2020 was one of the hottest years"***: "Climate Change Is Also a Racial Justice Problem," *The Washington Post*, https://www.washingtonpost.com/climate-solutions/2020/06/29/climate-change-racism.

13 ***"Coastal cities and towns"***: "Record-Breaking Atlantic Hurricane Season Draws to an End," National Oceanic and Atmospheric Administration, https://www.noaa.gov/media-release/record-breaking-atlantic-hurricane-season-draws-to-end.

13 ***"Coastal cities and towns"***: "North Florida Farmers Are Bracing for 'Devastating' Crop Losses after Hurricane Sally," *Tallahassee Democrat*, https://www.tallahassee.com/story/news/2020/09/23/farmers-brace-devastating-losses-because-hurricane-sally/5864481002.

14 ***"I was appointed as the Regional Adminstrator"***: EPA Region 4 oversees the following states: Alabama, Florida, Georgia, Kentucky, Mississippi, North Carolina, South Carolina and Tennessee.

CHAPTER 2

20 ***"Humans are throwing extra heat-trapping"***: "What is Global Warming?," Earth, https://www.facebook.com/watch/?v=1647869398820163.

21 ***"over 14.8 million acres"***: "Wildfires, Forest Fires around World in 2020," Anadolu Agency, https://www.aa.com.tr/en/environment/wildfires-forest-fires-around-world-in-2020/2088198.

21 ***"In June 2020, the number of fires"***: "Wildfires, Forest Fires around World in 2020."

21 ***"Greenland and Antarctica ice sheets"***: "Quick Facts on Ice Sheets," National Snow and Ice Data Center, https://nsidc.org/cryosphere/quickfacts/icesheets.html.

22 ***"Sometimes it's the poetic violence"***: "Strange Fruit" is a song performed by Billie Holiday, who first sang and recorded it in 1939. Written by teacher Abel Meeropol as a poem and published in 1937, it protested American racism, particularly the lynching of African Americans. Such lynchings had reached a peak in the South at the turn of the century, but continued there and in other regions of the United States. The great majority of victims were black. The song's lyrics are an extended metaphor linking a tree's fruit with lynching victims. Meeropol set it to music and, with his wife and the singer Laura Duncan, performed it as a protest song in New York City venues in the late 1930s, including Madison Square Garden.

24 ***"Unless we take action"***: "Climate Change 2021: The Physical Science Basis," The Intergovernmental Panel on Climate Change, https://www.ipcc.ch/report/ar6/wg1.

24 ***"At 1.5 degrees warming"***: "A Degree of Concern: Why Global Temperatures Matter," NASA, https://climate.nasa.gov/news/2865/a-degree-of-concern-why-global-temperatures-matter.

25 ***"Dr. Katherine Hayhoe explains"***: "So, Why is Two Degrees the Magic Number?," Global Weirding with Katharine Hayhoe, https://youtu.be/RhBBH8V3NPc.

25 ***"If there were a Divine Nine"***: The Divine Nine are the historically black Greek letter organizations. They are Alpha Phi Alpha Fraternity (1906), Alpha Kappa Alpha Sorority (1908), Kappa Alpha Psi Fraternity (1911), Omega Psi Phi Fraternity (1911), Delta Sigma Theta Sorority (1913), Phi Beta Sigma Fraternity (1914), Zeta Phi Beta Sorority (1920), Sigma Gamma Rho Sorority (1922) and Iota Phi Theta Fraternity (1963).

CHAPTER 3

35 ***"In 2020, fenceline communities"***: Fenceline communities are neighborhoods that are adjacent to pollution emitting facilities, such as petrochemical facilities and coal fired power plants.

35 ***"From the arrival of Tropical Storm Cristobal"***: Louisiana was hit with five named storms in 2020, Cristobal, Laura, Marco, Delta and Zeta, setting the record for the most landfalls in one season.

37 ***"The American Petroleum Institute (API)"***: "EGEB: Oil and Gas Lobbyist Are Trying to Stop Clean Energy with Facebook Ads," Electrek, https://electrek.co/2021/09/30/egeb-oil-and-gas-lobbyists-are-trying-to-stop-clean-energy-with-facebook-ads.

37 ***"In 2014, the Florida NAACP"***: "N.A.A.C.P. Tells Local Chapters: Don't Let Energy Industry Manipulate You," *The New York Times*, https://www.nytimes.com/2020/01/05/business/energy-environment/naacp-utility-donations.html.

37 ***"Then in 2017, a study from the NAACP"***: "Fumes Across the Fence-Line: The Health Impacts of Air Pollution from Oil & Gas Facilities on African American Communities," NAACP, https://naacp.org/resources/fumes-across-fence-line-health-impacts-air-pollution-oil-gas-facilities-african-american.

37 ***"As a result, African Americans"***: "Fumes Across the Fence-Line: The Health Impacts of Air Pollution from Oil & Gas Facilities on African American Communities."

37 ***"American Petroleum Institute (API) said nope"***: "Study Finds Pollution Puts African Americans at a Greater Risk of Getting Sick," Black Enterprise, https://www.blackenterprise.com/oil-gas-refineries-african-american-sick.

37 ***"API money poured into NAACP"***: "N.A.A.C.P. Tells Local Chapters: Don't Let Energy Industry Manipulate You."

38 ***"Led by award-winning"***: "Fossil Fueled Foolery 2.0," NAACP, https://naacp.org/resources/fossil-fueled-foolery-20.

CHAPTER 4

44 ***"petrochemical pollution"***: Petrochemicals are a chemical by-product of refining fossil fuel products such as natural gas and coal. The result creates one of the key materials needed to make plastic. In addition to contributing to the massive amount of plastics, the process of making petrochemicals releases an excessive amount of carbon dioxide into the atmosphere thereby contributing to the global warming crisis. According to the National Institute of Health, people who live around or near petrochemical facilities are disproportionally affected by health impacts such as shortness of breath, eye irritation, dizziness, cough, nose congestion, sore throat, phlegm and weakness from exposure to industrial air pollutants.

44 ***"Cancer Alley"***: Cancer Alley is the name given to the eighty-five mile stretch of land along the South Mississippi River between Baton Rouge and New Orleans, Louisiana. Beset with oil and gas refineries as well as petrochemical facilities, in the late 1980s residents and researchers began noting cancer clusters that appeared in small towns and communities surrounding the industries. The region suffers from a high rate of poverty and the community is majority African American and Indigenous. The term "Cancer Alley" became synonymous with environmental injustice in South Louisiana and also referred to these impacted populations.

45 ***"Goldman Environmental Prize for North America"***: "About the Goldman Prize," The Goldman Environmental Prize, https://www.goldmanprize.org/about. The Goldman Environmental Prize rewards grassroots environmental heroes from six regions: Africa, Asia, Europe, Islands and Island Nations, North America, and South and Central America. "The Prize recognizes individuals for sustained and significant efforts to protect and enhance the natural environment, often at great personal risk. The Goldman Prize views 'grassroots' leaders as those involved in local efforts, where positive change is created through community or citizen participation."

47 ***"Less than seventy-two hours after Hurricane Ida"***: "Hurricanes Ida and Nicholas | Update #19," U.S. Department of Energy, www.energy.gov/sites/default/files/2021-09/TLP-WHITE_DOE%20Situation%20Update_Hurricane%20Ida_19.pdf.

49 ***"Shintech Louisiana in Plaquemine reported"***: "Hurricanes Ida and Nicholas | Update #19."

CHAPTER 5

55 ***"It was my responsibility"***: Region 4 includes the southeastern states of Alabama, Florida, Georgia, Kentucky, North Carolina, South Carolina, Mississippi and Tennessee. Region 4 is home to the six federally recognized tribes: Catawba Indian Nation, Eastern Band of Cherokee Indians, Miccosukee Tribe of Indians of Florida, Mississippi Band of Choctaw Indians, Poarch Band of Creek Indians and Seminole Tribe of Florida. Region 4 is also home to a number of state and locally recognized Indigenous people.

56 ***"the Federal Housing Administration (FHA) refused"***: "How a New Deal Housing Program Enforced Segregation," HISTORY, https://www.history.com/news/housing-segregation-new-deal-program.

56 ***"the Federal Housing Administration (FHA) refused"***: *The Color of Law*, Liveright Publishing, 2017.

56 ***"The FHA's Underwriting Manual"***: *The Color of Law*, Liveright Publishing, 2017.

56 ***"Today, African Americans carry the weight"***: "Disparities in the Impact of Air Population," American Lung Association, https://www.lung.org/clean-air/outdoors/who-is-at-risk/disparities.

56 ***"Not only does transportation pollution"***: "Living near Highways and Air Pollution," American Lung Association, https://www.lung.org/clean-air/outdoors/who-is-at-risk/highways.

57 ***"Although the Roosevelt administration"***: *The Black Cabinet: The Untold Story of African Americans and Politics During the Age of Roosevelt*, Grove Press, 2020. Also known as the Black Brain Trust, President Roosevelt appointed forty-five notable African American leaders to serve as advisors to his administration. Members included group leader Dr. Mary McLeod Bethune (Founder of Bethune-Cookman College and good friend of First Lady Eleanor Roosevelt), Dr. Robert Clifton Weaver (He later became the first African American appointed to a US Cabinet level position. He served as Secretary of Housing and Urban Development (HUD) under President Lyndon B. Johnson) and Mr. Lawrence A. Oxley (one of the top social program specialists in the United States. An active civic member, Mr. Lawrence served as the National President of Omega Psi Phi Fraternity, Inc. He wrote the guide outlining how Black American youth could become federal government employees.).

57 ***"Although the Roosevelt administration"***: "How a New Deal Housing Program Enforced Segregation," HISTORY, https://www.history.com/news/housing-segregation-new-deal-program.

57 ***"The Great Depression"***: "Lynching in America: Confronting the Legacy of Racial Terror," Equal Justice Initiative, https://eji.org/reports/lynching-in-america.

57 ***"The Great Depression"***: "Great Migration," HISTORY, https://www.history.com/topics/black-history/great-migration-video. In 1900, Black Americans accounted for nine out of every ten Black people in the US. By 1970, the south accounted for only 20 percent of Black Americans in the country. Almost six million Black Americans migrated to the North from the Southeast United States by 1960.

57 ***"In 1933, The Home Owners' Loan Act"***: The Home Owners' Loan Act of 1933, also known as the Homeowners Refinancing Act, sought to "refinance home mortgages [and] to extend relief to the owners of homes who occupy them who are unable to amortize their debt elsewhere." Enacted during the Great Depression, the law established the Home Owners' Loan Corporation (HOLC). This emergency federal agency provided mortgage assistance to homeowners by lending low-interest money, refinancing mortgages, and originating new mortgages. HOLC issued government insured bonds to local lenders in exchange for delinquent mortgages in their portfolios. In addition to alleviating the Great Depression, the Home Owners' Loan Act of 1933 forever changed America's mortgage market. Before the HOLC, most home loans had three to five years terms, with high interest rates and a closing principal, or "balloon," payment. HOLC established and normalized a fifteen-year amortizing loan, which allowed homeowners to pay off their mortgages in monthly installments over many years with the principal reduced over time. This change in mortgage finance would ultimately lead to the modern 30-year fixed-rate mortgage.

58 ***"However, redlining forms"***: "Mapping Inequality," Digital Scholarship Lab, https://dsl.richmond.edu/panorama/redlining/#loc=13/33.534/NaN&city=birmingham-al&area=D1&adimage=4/80.134/-165.85.

58 ***"Exclusionary zoning and suburban covenants"***: "Addressing Community Concerns: How Environmental Justice Relates to Land Use Planning and Zoning," *National Academy of Public Administration*, https://www.epa.gov/sites/default/files/2015-02/documents/napa-land-use-zoning-63003.pdf.

59 ***"HOLC created "Residential Security" maps"***: "HOLC 'Redlining' Maps: The Persistent Structure of Segregation and Economic Inequality," NCRC, https://ncrc.org/holc. Cities redlined by the federal government: **Alabama:** Birmingham, Mobile, Montgomery **Arizona:** Phoenix, **Arkansas:** Little Rock, **California:** Fresno, Los Angeles, Oakland, Sacramento, San Diego, San Francisco, San Jose, Stockton, **Colorado:** Denver, Pueblo, **Connecticut:** Hartford, New Britain, New Haven, Stamford, Darien, and New Canaan, Waterbury, **Florida:** Jacksonville, Miami, St. Petersburg, Tampa, **Georgia:** Atlanta, Augusta, Columbus, Macon, Savannah, **Illinois:** Aurora, Chicago, Decatur, East St. Louis, Joliet, Peoria, Rockford, Springfield, **Indiana:** Evansville, Fort Wayne, Indianapolis, Lake Co. Gary, Muncie, South Bend, Terre Haute, **Iowa:** Council Bluffs, Davenport, Des Moines, Dubuque, Sioux City, Waterloo, **Kansas:** Topeka. Wichita, **Kentucky:** Covington, Lexington, Louisville, **Louisiana:** New Orleans, Shreveport, **Maryland:** Baltimore, **Massachusetts:** Arlington, Belmont, Boston, Braintree, Brockton, Brookline, Cambridge,

Chelsea, Dedham, Everett, Haverhill, Holyoke Chicopee, Lexington, Malden, Medford, Melrose, Milton, Needham, Newton, Quincy, Revere, Saugus, Somerville, Waltham, Watertown, Winchester, Winthrop, **Michigan:** Battle Creek, Bay City, Detroit, Flint, Grand Rapids, Jackson, Kalamazoo, Lansing, Muskegon, Pontiac, Saginaw, **Minnesota:** Duluth, Minneapolis, Rochester, St. Paul, **Mississippi**: Jackson, **Missouri:** Greater Kansas City, Springfield, St. Joseph, St. Louis, **Nebraska**: Lincoln, Omaha, **New Hampshire**: Manchester, **New Jersey**: Atlantic City, Bergen Co., Camden, Essex Co., Hudson Co., Trenton, Union Co., **New York:** Albany, Binghamton-Johnson City, Bronx, Brooklyn, Buffalo, Elmira, Lower Westchester Co., Manhattan, Niagara Falls, Poughkeepsie, Queens, Rochester, Schenectady, Staten Island, Syracuse, Troy, Utica, **North Carolina**: Asheville, Charlotte, Durham, Greensboro, Winston-Salem, **Ohio**: Akron, Canton, Cleveland, Columbus, Dayton, Hamilton, Lima, Lorain, Portsmouth, Springfield, Toledo, Warren, Youngstown, **Oklahoma**: Oklahoma City, Tulsa, Oregon, Portland, **Pennsylvania**: Altoona, Bethlehem, Chester, Erie, Harrisburg, Johnstown, Lancaster, New Castle, **Philadelphia**: Pittsburgh, Wilkes-Barre, York, **Rhode Island**: Pawtucket & Central Falls, Providence, Woonsocket, **South Carolina**: Columbia, **Tennessee:** Chattanooga, Knoxville, Memphis, Nashville, **Texas:** Amarillo, Austin, Beaumont, Dallas, El Paso, Fort Worth, Galveston, Houston, Port Arthur, San Antonio, Waco, **Utah:** Ogden, Salt Lake City, **Virginia**: Lynchburg, Newport News, Norfolk, Richmond, Roanoke, **Washington:** Seattle, Spokane, Tacoma, **West Virginia**: Charleston, Huntington, Wheeling, **Wisconsin**: Kenosha, Madison, Milwaukee Co., Oshkosh, Racine.

59 ***"One of the cities mapped"***: "Not Even Past: Social Vulnerability and the Legacy of Redlining," Digital Scholarship Lab, https://dsl.richmond.edu/socialvulnerability/map/#loc=11/33.591/-86.717&city=birmingham-al.

59 ***"According to the map"***: "Mapping Inequality," Digital Scholarship Lab, https://dsl.richmond.edu/panorama/redlining/#loc=13/33.534/NaN&city=birmingham-al&area=D1&adimage=4/80.134/-165.85.

59 ***"The negatives identified"***: "Mapping Inequality."

60 ***"Consent decrees"***: A Consent Decree is a legal settlement between the United States Department of Justice and an entity (industry, city government, state, etc.) that has committed violations of a federal environmental regulation.

68 ***"A strong community revitalization plan"***: "Northern Birmingham Revitalization Action Plan," *Northern Birmingham Community Coalition,* https://archive.epa.gov/epa/sites/production/files/2016-05/documents/r4_north_birmingham_nbcc_action_plan_2015_01_16.pdf.

69 ***"In exchange for the money"***: "Former State Legislator Sentenced to 33 Months in Prison for Accepting Bribes," The United States Department of Justice, https://www.justice.gov/usao-ndal/pr/former-state-legislator-sentenced-33-months-prison-accepting-bribes.

70 ***"In 2018, the regional administrator"***: "Several Charges Dropped for Trey Glenn in EPA Scandal," Advance Local, https://www.al.com/news/birmingham/2019/02/several-charges-dropped-for-trey-glenn-in-epa-scandal.html.

70 ***"Alabama has twelve sites"***: "Superfund: National Priorities List (NL)," United States Environmental Protection Agency, https://www.epa.gov/superfund/superfund-national-priorities-list-npl.

75 ***"Turns out, the thing you make"***: "What Is Coke?," Clean Air Council, https://pacokeovens.org/what-is-coke.

CHAPTER 6

81 ***"I was attending the annual Congressional Black Caucus"***: Steve Pruitt passed away January 15, 2020.

82 ***"As the former President"***: ArkLaMiss—The Arkansas, Louisiana, Mississippi Delta region. The area that all three states meet.

82 ***"President Clinton"***: Doe's Eat Place is a famous small family owned steak restaurant in Greenville, Mississippi. It sits in a little shotgun house on Nelson Street. Bill Clinton was known for traveling from Arkansas, across the Mississippi River Bridge and to downtown Greenville just to eat at Doe's Eat Place.

84 ***"An organization, usually nonprofit"***: A Trust is a legal mechanism for allowing one party (person or organization) to control the assets of another for the benefit of a third party. For example, parents can establish a trust fund for a child that is controlled by another adult until the child comes of the age established in the trust.

85 ***"There are currently over two hundred"***: "Community Land Trust Directory," Schumacher Center, https://centerforneweconomics.org/apply/community-land-trust-program/directory.

85 ***"Northeast Farmers of Color Land Trust"***: From "Home Page," Northeast Farmers of Color Land Trust, https://nefoclandtrust.org. An organization that focuses on restorative and reparations for BIPOC people through farming.

CHAPTER 7

93 ***"Next to fossil fuels"***: "Global Food System Emissions Could Preclude Achieving the 1.5° and 2°C Climate Change Targets," *Science*, https://science.sciencemag.org/content/370/6517/705.

95 ***"According to the FBI"***: "Grocery Workers Have Borne the Brunt of the Pandemic. Now Supermarket Shootings Are on the Rise.," *The Washington Post*, https://www.washingtonpost.com/nation/2021/09/25/us-grocery-store-shootings.

95 ***"In 2021 alone, we saw the deaths"***: "Grocery Workers Have Borne the Brunt of the Pandemic. Now Supermarket Shootings Are on the Rise."

96 ***"The Supplemental Nutrition Assistance Program"***: From "Supplemental Nutrition Assistance Program (SNAP)," USDA, https://www.fns.usda.gov/snap/supplemental-nutrition-assistance-program.

96 ***"Most people know it as"***: The original Food Stamp Program was started in 1939 and ended in 1942. It kicked off again in 1961 after soon-to-be President Kennedy made a campaign promise in West Virginia. The first recipients of the Food Stamp Program were Mr. and Mrs. Alderson Muncy of Paynesville, West Virginia and the program soon expanded from one county to eight states.

98 ***"In 2016, the Pew Research Center"***: "The New Food Fights: U.S. Public Divides over Food Science," Pew Research Center, https://www.pewresearch.org/science/2016/12/01/the-new-food-fights.

98 ***"In fact, a third of Black Americans"***: "Nearly One in Four in U.S. Have Cut Back on Eating Meat," Gallup, https://news.gallup.com/poll/282779/nearly-one-four-cut-back-eating-meat.aspx.

98 ***"While the United States remains"***: "Nearly One in Four in U.S. Have Cut Back on Eating Meat."

98 ***"Even with the knowledge of these polls"***: "The Vegan Race Wars: How the Mainstream Ignores Vegans of Color," Thrillist, https://www.thrillist.com/eat/nation/vegan-race-wars-white-veganism.

102 ***"I was sitting at Ted Climate Countdown"***: From "Deep South Center for Environmental Justice," Deep South Center for Environmental Justice, https://www.dscej.org.

102 ***"WE ACT for Environmental Justice"***: From "WE ACT for Environmental Justice," WE ACT for Environmental Justice, https://www.weact.org.

103 ***"According to the EPA"***: "Inventory of U.S. Greenhouse Gas Emissions and Sinks," United States Environmental Protection Agency, https://www.epa.gov/sites/default/files/2021-04/documents/us-ghg-inventory-2021-main-text.pdf?VersionId=yu89kg1O2qP754CdR8Qmyn4RRWc5iodZ.

103 ***"Livestock scientists found"***: "How Seaweed Could Reduce Cow's Methane Emissions by up to 90%," mindbodygreen, https://www.mindbodygreen.com/articles/seaweed-for-cows-methane.

104 ***"When incorporated into the feedstock"***: "How Seaweed Could Reduce Cow's Methane Emissions by up to 90%."

104 ***"The scientists"***: "How Seaweed Could Reduce Cow's Methane Emissions by up to 90%."

CHAPTER 8

107 ***"Sixty percent days"***: Pursuant to Mississippi Code 37-13-63, all public schools in the state shall be kept in session for at least 180 days in each scholastic year. For the purpose of determining a full school day, teachers and students must be present for at least sixty-three percent (63%) of the instructional day, as fixed by the local school board. Mississippi Law, https://www.sos.ms.gov/communications-publications/mississippi-law.

109 ***"As integration made its way"***: White flight is the large scale migration of white people from an area that is becoming more ethnically diverse.

109 ***"By the time I entered the public school system"***: Greenville High School and T.L. Weston High School were public schools. Washington School (1969), Greenville Christian School (1969) and St. Joseph High School (1888) are private or parochial schools. Sacred Heart School was the Black Catholic school established in 1920 and closed in 1970.

110 ***"Coleman became a junior high school"***: One additional school, E.E. Bass, was also used as a junior high school during the integration period. Built in 1916, it was the Mississippi Delta's first and oldest high school. E.E. Bass was used as an integration school to combine Black and white students before feeding into the larger integrated high school system. The age and structure of the school would not accommodate costly building, wide upgrades such as central air and heating. E.E. Bass was soon thereafter transitioned into a community building and remains as a historic arts center in the community today.

110 ***"A study in the Nature Human Behaviour journal"***: (Flavelle, 2020)

111 ***"Public housing developments"***: Impervious surfaces are areas covered with material that does not allow for the free flow and infiltration of storm water into the ground.

112 ***"A study similar to the Nature Human Behavior"***: (E. C. Merem, Vol. 6 No. 1, 2016).

112 ***"When power plants release emissions"***: Power plant emissions include sulfur dioxide (SO_2), nitrogen oxides (No_x), particulate matter (PM) and other hazardous air pollutants that are subject to environmental regulations.

113 ***"susceptibility to lead poisoning"***: "Lead Poisoning Reveals Environmental Racism in the US," DW Akademie, https://www.dw.com/en/lead-poisoning-reveals-environmental-racism-in-the-us/a-53335395.

114 ***"Historically Black colleges and universities"***: HBCU—Historically Black Colleges and Universities.

114 ***"minority-serving institutions"***: MSI—Minority-Serving Institutions.

115 ***"In 2021, the New York Times"***: "The Tragedy of America's Rural Schools," *The New York Times*, https://www.nytimes.com/2021/09/07/magazine/rural-public-education.html.

115 ***"Students may grow distracted"***: "The Tragedy of America's Rural Schools."

CHAPTER 9

120 ***"The June 2020 Time magazine"***: "The Story Behind TIME's George Floyd Cover," *Time*, https://time.com/5847667/story-behind-george-floyd-time-cover.

123 ***"In 2013, the New York Times"***: "Weather and Violence," *The New York Times*, https://www.nytimes.com/2013/09/01/opinion/sunday/weather-and-violence.html.

123 ***"Our leadership must call"***: "Weather and Violence."

123 ***"More recent studies show"***: "Climate Change and Crimes in Cities," Igarapé Institute, https://igarape.org.br/wp-content/uploads/2021/07/Climate-change-and-crime-in-cities.pdf.

123 ***"A Brazilian study"***: "Climate Change and Crimes in Cities."

123 ***"These two factors"***: "Climate Change and Crimes in Cities."

123 ***"In a follow up 2018 piece"***: "A Rise in Murder? Let's Talk About the Weather," *The New York Times*, https://www.nytimes.com/2018/09/21/upshot/a-rise-in-murder-lets-talk-about-the-weather.html.

123 ***"In cities like Chicago and New York"***: "A Rise in Murder? Let's Talk About the Weather."

CHAPTER 10

131 ***"Candyman"***: *Candyman* is a 1992 American horror film set in Chicago's predominantly Black Cabrini Green housing projects and features a Black lead actor as the villain. Set on the theme of urban legends, it is the horrifying story of a student who says the name "Candyman" five times in a mirror. The urban legend becomes real in the form of a Black man with a hook for a hand that terrorizes and murders victims throughout the projects. The film was updated in 2021 and even further explores the racial trauma that created *Candyman* in the first place. The film tackles the strength of myths and urban legends in the Black community and, despite it being fiction, verifies many of the strong beliefs in Black American myths and urban legends as well as how they are passed from generation to generation.

132 ***"Anne Moody's Mississippi"***: *Coming of Age in Mississippi: The Classic Autobiography of Growing Up in the Rural South*, Lulu.com, 1968. In Anne Moody's memoir, Anne's mother explains Emmitt Till's death by saying that an evil spirit killed him.

133 ***"At the end of the movie"***: "White Females at Highest Risk in Horror Films," Time, https://time.com/3547214/horror-films-who-dies-first.

133 ***"Rev. Ralph Warnock"***: Rev. Ralph Warnock is the senior pastor of historic Ebenezer Baptist Church in Atlanta, Georgia, where Rev. Dr. Martin Luther King Jr. and his father both served as senior pastor. Rev. Warnock is the first Black Senator for the state of Georgia.

134 ***"By December 2020, state legislatures"***: "Voting Laws Roundup: March 2021," Brennan Center for Justice, https://www.brennancenter.org/our-work/research-reports/voting-laws-roundup-march-2021.

134 ***"By July 2021, seventeen states"***: "Report: Republican-Led State Legislatures Pass Dozens of Restrictive Voting Laws in 2021," U.S. News & World Report, https://www.usnews.com/news/best-states/articles/2021-07-02/17-states-have-passed-restrictive-voting-laws-this-year-report-says.

135 ***"She went on ABC's The View"***: "Stacey Abrams Addresses the Possibility of Joining Biden Administration," ABC News, https://abcnews.go.com/theview/video/stacey-abrams-addresses-possibility-joining-biden-administration-74175132.

136 ***"Study after study has shown"***: "Unequal Impact: The Deep Links between Racism and Climate Change," YaleEnvironment360, https://e360.yale.edu/features/unequal-impact-the-deep-links-between-inequality-and-climate-change.

136 ***"The people most impacted"***: "Chapter 2: Climate Change and Energy Issues," Pew Research Center, https://www.pewresearch.org/science/2015/07/01/chapter-2-climate-change-and-energy-issues.

136 ***"He claimed that he wanted 'immaculate air and water'"***: The Paris Climate Accord (also called the Paris Climate Agreement) is an international treaty on climate change. It was initially signed by 196 countries, including the United States of America, in 2015. Under the Trump administration, the United States pulled out of the Paris Climate Accord thereby ceasing all

US actions on climate change and becoming the first nation in the world to formerly withdraw from the international agreement.

137 ***"This was the same man"***: "Trump Calls for Review of Water Efficiency Standards, Saying People Flush the Toilet '10 Times, 15 Times'," *The Hill*, https://thehill.com/policy/energy-environment/473450-trump-calls-for-review-of-water-efficiency-standards-saying-people.

138 ***"In 1980, Audre Lorde"***: *Sister Outsider: Essays and Speeches*, Crossing Press, 1984.

139 ***"In 2016, only 45 percent of white women"***: "An Examination of the 2016 Electorate, Based on Validated Voters," Pew Research Center, https://www.pewresearch.org/politics/2018/08/09/an-examination-of-the-2016-electorate-based-on-validated-voters.

139 ***"While a lower number of white women"***: "This Is How Women Voters Decided the 2020 Election," MSNBC, https://www.nbcnews.com/know-your-value/feature/how-women-voters-decided-2020-election-ncna1247746.

140 ***"This is true in areas"***: "Could the Voting Rights Fight Hinder Climate and Energy Policies?," *POLITICO*, https://www.politico.com/news/2021/07/20/voting-rights-climate-change-energy-justice-500288.

140 ***"Lovecraft Country"***: The HBO series, *Lovecraft Country*, was written by Misha Green and based on the 2016 novel of the same name by Matt Ruff. Both the novel and the television series builds upon the history of famous horror author Howard Phillip Lovecraft. Born in 1890, H.P. Lovecraft was known for creating fantasy and horrific worlds with systems based in race and class.

141 ***"Lovecraft is so crosscutting"***: "Candyman: The Official Companion Guide," Langston League, https://langstonleaguellc.squarespace.com/popculturepd.

141 ***"When Texas Governor Abbott demanded"***: "Inside the Texas Democratic Walkout That Derailed Senate Bill 7," *Texas Monthly*, https://www.texasmonthly.com/news-politics/texas-democrats-walkout-senate-bill-7.

141 ***"For constituents who live"***: "Climate Change May Have Worsened Deadly Texas Cold Wave, New Study Suggests," *The Washington Post*, https://www.washingtonpost.com/weather/2021/09/03/climate-change-arctic-texas-cold.

142 ***"Like the powerhouse leading women"***: From "Higher Heights for America," Higher Heights for America, https://www.higherheightsforamerica.org.

142 ***"The Time 100 list"***: "How We Chose the 2020 TIME100," *Time*, https://time.com/5891201/how-we-chose-time100-2020.

142 ***"This is Lovecraft Country"***: "Lovecraft Country and Wynonna Earp Give Women the Space to Be Bold," *ELLE*, https://www.elle.com/culture/movies-tv/a33648915/lovecraft-country-wynonna-earp-women-in-sci-fi-fantasy.

142 ***"Strategize like Ruby"***: In the *Lovecraft Country* series, Ruby, unbeknownst to her, drank a potion that turned her into a white woman for a short period of time. She turned it into an opportunity to spy for the benefit of her family. Hippolyta was a scientist who was withheld from certain opportunities because she was a wife and a mother. She time traveled and used facts and science to save her daughter. Letti, Ruby's sister, was an all-around badass, bold and could best be described as Indiana Jones's little sister. Her bravery gave her the strength to walk through the middle of the Tulsa Massacre fire bombing, unscathed, to save the family's history and future.

144 ***"Minority people are more likely to vote":*** "Which Racial/Ethnic Groups Care Most About Climate Change?," Yale Program on Climate Change Communication, https://climatecommunication.yale.edu/publications/race-and-climate-change.

CHAPTER 11

149 ***"All associated with a sinful world"***: "For the creation was subjected to futility, not willingly, but because of Him who subjected it in hope; because the creation itself also will be delivered from the bondage of corruption into the glorious liberty of the children of God. For we know that the whole creation groans and labors with birth pangs together until now" (Romans 8:20–22 NKJV).

152 ***"To better understand how influential faith is"***: "5 Facts about the Religious Lives of African Americans," Pew Research Center, https://www.pewresearch.org/fact-tank/2018/02/07/5-facts-about-the-religious-lives-of-african-americans.

CHAPTER 12

164 ***"Studies have shown"***: "Which Racial/Ethnic Groups Care Most About Climate Change?," Yale Program on Climate Change Communication, https://climatecommunication.yale.edu/publications/race-and-climate-change.

165 ***"Leaders on the council"***: Dr. Beverly Wright is the founder and director of the Deep South Center for Environmental Justice, Dr. Robert Bullard is called the "father of Environmental Justice" and has written nineteen books on the subject. Catherine Flowers is a MacArthur Fellow and author of the book, *Waste: One Woman's Fight Against America's Dirty Secret*, and former State Representative Harold Mitchell is one of the main thought leaders and creators of the ReGenesis program which leveraged a $300,000 EPA workforce development grant into a $30,000,000 neighborhood revitalization and environmental sustainability project.